CONCISE COLOR GUIDES

Aircraft
World Wars I & II

General Editor
Jeff Daniels

Longmeadow Press

This 1988 edition is published by
Longmeadow Press
201 High Ridge Road
Stamford CT 06904

ISBN 0 681 40431 0
Printed in Italy

0987654

Contents

Introduction
by Jeff Daniels 4

World War I
Aircraft 16

World War II
Aircraft 92

Index 236

Introduction

The aircraft described in this book cover a huge range of developments in both scope and ability. In fact only 31 years span 1914 to 1945, yet in that time the military aeroplane went from being a flimsy machine which could barely fly at all, spent its time 'scouting' and dropping tiny improvised bombs, to become one of the first 885km/h (550mph) jet fighters as well as a bomber capable of dropping the atom bombs on Japan. Whole areas of specialised interest for the military aeroplane designer emerged. The division between fighter and bomber was the most obvious, but there were many others. The ground between fighter and bomber was occupied

De Havilland Mosquito

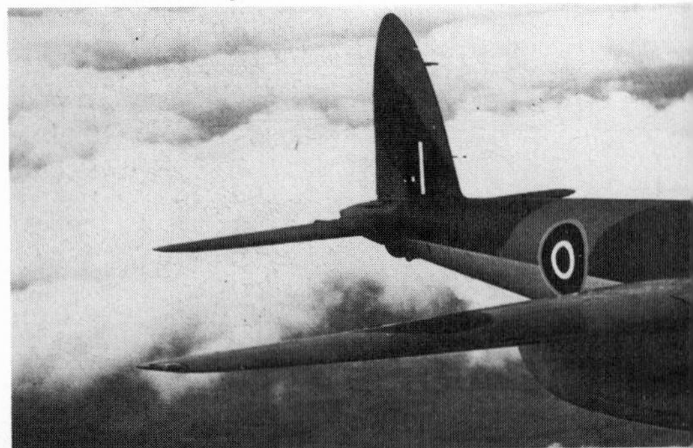

by a number of aeroplanes, like the de Havilland Mosquito, which were capable of being either. The fighter designer might work on the classic single-engined, single-seater, or look at a heavier twin-engined aeroplane to carry the radar needed for night-fighting. The bomber designer had to choose between the versatile 'twin' and the heavy four-engined design. There were aeroplanes with a special 'line' in tank-busting, such as Russia's Ilyushin Il-2, the *Shturmovik*; there were reconnaissance aircraft; aircraft like the Westland Lysander which operated close to the army and dropped agents behind enemy lines; flying boats and long-range maritime patrol aircraft; there were also a host of designs adapted to the special problems of operating from aircraft carriers at sea.

The almost carefree days of 1914 did not last long. Pilots - and others - soon learned that the pressures of war act to increase the pace of development - even if by trial and error. Engines quickly became more powerful and reliable, airframes tougher, control systems more capable. At first army commanders thought the aeroplane was the perfect way to fly across and find out what the enemy was up to. Some of the earliest military aircraft, like the R.A.F. FE2, were laid out specially to give the observer the clearest possible field of view. It quickly emerged that such 'pusher' designs were slow and clumsy compared with the first 'tractor' fighters like the Fokker E.III which tried to shoot them out of the sky. By 1915, too, the ordnance factories had worked out how to make guns which could shoot very effectively at slow-moving aeroplanes.

By 1916, therefore, a clearer picture was already emerging. The early 'scout' had given way to the reconnaissance aeroplane, faster and properly armed, and often equipped with cameras as well as the crew's sharp eyes. The 'recce' types were often escorted by the single-seater fighters which each warring nation was striving to develop: Britain the Bristol Scout and Sopwith Pup, France the Nieuport and the Spad, Germany the Fokker, the Albatros and the Pfalz. The Germans were the first to discover a safe way of making a machine-gun shoot between the blades of a turning propeller, so that the pilot could aim simply by pointing his whole aeroplane in the right direction, but their advantage did not last long: and more advanced fighters like the Sopwith Camel and the S.E.5 slugged it out with the Fokker D.VIIs over the trenches. It had already been realised in essence that whatever the importance of reconnaissance, the first issue to be decided in any air war was who controlled the sky – and that was work for the fighters.

At the same time, other specialised types were making their appearance. It was to be a long time before the strategic potential of bombing was fully realised, but the idea of striking at enemy industrial and communications centres was an appealing one – and so was that of building bigger and more imposing aeroplanes. Thus, before World War I was over, machines like the Handley Page O/400 and the Gotha G.V had taken the concept of bombing well beyond dropping glorified grenades on the front-line trenches. That did not mean the trenches had a quieter time, for the aircraft designers had already adapted their fighters to take armour plate and special armament for the task of attacking small targets from low level. Designs like the Sopwith Dolphin were the forerunners of a later generation of specialised ground-attack aeroplanes.

Once World War I was over, interest in aircraft development seemed almost to stop. It is fair to say that over the 12 years from 1918 to 1930, military aircraft performance and ability showed far less improvement, relatively speaking, than in the four years 1914–18. Had it not been for the intense competition for the Schneider Trophy, it is doubtful whether aircraft designers in Britain at least would have been allowed to think about more powerful engines and high-performance monoplanes until it was too late to prepare the RAF for the outbreak of the war against Germany in 1939.

The switch from the biplane to the monoplane did not come easily. There were plenty of senior air force officers in every major country who were steeped in the lessons of World War I and saw the biplane as the best solution for almost everything: for fighters because it made them more manoeuvrable, for bombers because it gave them greater lifting power – and for both, because it meant slow take-off and landing speeds which meant aeroplanes could be operated from small grass airfields. It was only in the 1930s that the biplane proved almost to have limited itself: no matter how much extra power was poured in, no biplane figher ever exceeded 480km/h (300mph), speed which quickly became commonplace for the monoplanes that followed.

By 1939 the designers' efforts had been produced and by 1940 the principal opponents stood face to face across the English Channel. The Supermarine Spitfire and Hawker Hurricane had to fight through the protective screens of Messerschmitt Bf109s and 110s to reach the German bombers – mainly, in those early days, the Dornier Do17 and the Heinkel He111. It said something for the pace of develop-

Hawker Hurricanes

ment in air warfare that the lessons learned by the Germans during the Spanish Civil War availed them nothing during the Battle of Britain. A shift of scene and a different kind of opponent were enough to throw the whole question of fighting technique wide open once more. It was clear after 1940 that the single-engined, single-seat fighter still reigned supreme in any battle for air superiority. Fighters with two engines, or two seats, like the Messerschmitt Bf110 or the Boulton Paul Defiant, were at an immediate disadvantage.

Britain's bombing counter-offensive started very early in the war with a trio of twin-engined bombers – Armstrong Whitworth Whitley, Handley Page Hampden and Vickers Wellington – which were fairly direct counterparts of the German Dornier and Heinkel. However, Britain's factories were also making ready three more formidable four-engined designs, the Avro Lancaster, Short Stirling and Handley Page Halifax. By 1942 these machines were carrying

Short Stirling

up to 10 tonnes (10 tons) of bombs on night-time raids over Germany; yet the Germans never responded with a 'proper' four-engined bomber of their own.

World War II was notable for the widening of the scope of military aviation in every sense. In addition to the best-known and widely used British and German designs, the years 1941–45 in particular saw many capable aeroplanes emerge from American, Russian, Japanese and Italian factories. It was a period which saw the emergence of some Russian design teams whose acronyms have since become so familiar: MiG and Yak built tough and successful single-seat fighters, while Ilyushin and Tupolev concentrated on heavier types. As for the Italians, those who fought in the Mediterranean never under-estimated the Fiat and Macchi fighters, nor indeed the ubiquitous Savoia Marchetti S.M. 79 three-engined bomber. At the same time, aeroplanes operated in ways hardly dreamed of during the previous war: in particular, from the heaving decks of aircraft carriers, and on long-range maritime patrols to protect convoys against submarines.

Because the Pacific War was fought almost entirely over water, many of the best-known American and Japanese aircraft were carrier-borne – either fighters like the Grumman F4F Wildcat and F6F Hellcat or the Mitsubishi A6M5 'Zeke' (almost always known in practice as the Zero); or torpedo bombers like the Grumman Avenger and the Aichi

D3A1 'Val', which caused such havoc at Pearl Harbour. Yet both nations also produced formidable land-based designs, and in the European theatre the Boeing B-17 and Consolidated B-24 became familiar sights as they went out to raid Germany by day, escorted by the Republic P-47 Thunderbolts and North American P-51 Mustangs. The Americans also contributed three high-performance twin-engined medium bombers, the Douglas A-20 Havoc, the North American B-25 Mitchell and the Martin B-26 Marauder; but the highest-performance bomber design of all was the de Havilland Mosquito.

The Mosquito, along with Bristol's Beaufighter, became the key design in a new type of fighting – intercepting night bombers. The invention of radar indicated the need for a fighter big enough to carry both the radar set and its operator, and in this case two engines really were the answer. German thinking ran along the same lines, as we can see in the formidable Heinkel He 219 night-fighter that tried to stem the British bomber streams in 1944–45.

The importance of convoy patrol work over the Atlantic should never be under-estimated, and for much of the war this work was undertaken by two stalwart flying-boats, the Short Sunderland and the Consolidated PBY Catalina. By 1944 however many patrols were also flown by adapted four-engined bombers, especially the Consolidated B-24 Liberator, and indeed it was the building of so many concrete runways during the war which in the end

Gloster Meteor F.III

signalled the eventual demise of the flying-boat.

Once the war had started, the Germans seemed to prefer to develop their existing designs rather than embarking on completely new ones. Thus the only all-new fighter to appear on their side until 1944 was the Focke-Wulf Fw 190, while the British (and especially the Hawker design team) produced ever more powerful single-seaters. However, the Germans were working hard on their first jet combat aircraft, the first and most famous of which was the twin-engined Messerschmitt Me 262 fighter. By the end of the war, they had also put into service the first operational jet bomber, the Arado Ar 234. Meanwhile the British were following their own jet develop-

ment programme, and the Gloster Meteor was in service in time to combat the 'flying bomb' menace. Indeed, the advent of the jet engine signalled the start of a new age in military design, and by the end of World War II it might well be said that virtually all the aircraft which had fought in it were already obsolete.

Author's Note

Each illustration is accompanied by data for the model or variant listed in the title. Some of the aircraft are also illustrated with three views of similar models or slightly different variants. We have given the data for the most typical models of the aircraft. However many of the aircraft did appear with different engines and armament.

Airco D.H.2

The Airco D.H.2 was a smaller version of the pusher scout D.H.1 designed by Geoffrey de Havilland. It was tricky to fly as the pilot had to aim and fire the gun while flying the aircraft. However it was a very manoeuvrable aircraft and was able to master the Fokker E.III. Altogether 450 were built and about 300 served in France.

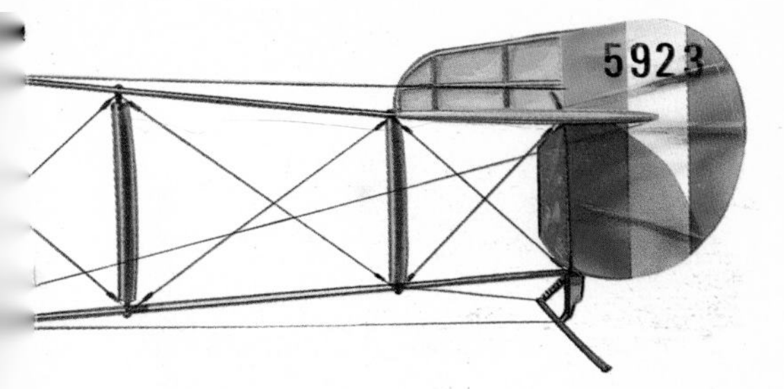

Country of Origin: Great Britain
Type: fighter
Accommodation: one
Armament: single or twin 8mm (.303in) Lewis machine-guns in nose, dorsal and tail cockpits plus single firing down through fuselage floor.
Power Plant: one 100hp Gnome Monosoupape rotary
Performance: maximum speed 150km/h (93mph); service ceiling 4,420m (14,500ft); endurance 2¾ hours
Weight: take-off 654kg (1,441lb)
Dimensions: span 8.6m (28ft 3in); length 7.7m (25ft 2.5in); wing area 23.1m² (249 sq ft)
Year entered service: 1915

Albatros D.III

The Albatros D.III was one of the great fighting scouts of World War I. It helped the Imperial German Air Service achieve air superiority in the first half of 1917. Some 2,000 were built and it was flown by the Richthofen 'circus'.

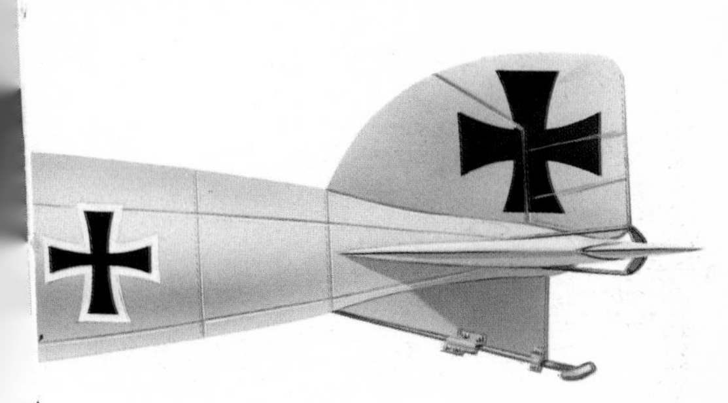

Country of Origin: Germany
Type: fighter
Accommodation: one
Armament: two 7.92mm Spandau machine-guns firing forward
Power Plant: one 160/175hp Mercedes D.IIIa water-cooled in-line
Performance: maximum speed 175km/h (109mph) at 1,000m (3,280ft); service ceiling 5,500m (18,045ft); endurance 2 hours
Weight: take-off 886kg (1,953lb)
Dimensions: span 9.05m (29ft 8.33in); length 7.3m (24ft .5in); wing area 20.5m^2 (220.7sq ft)
Year entered service: 1917

Albatros D.Va

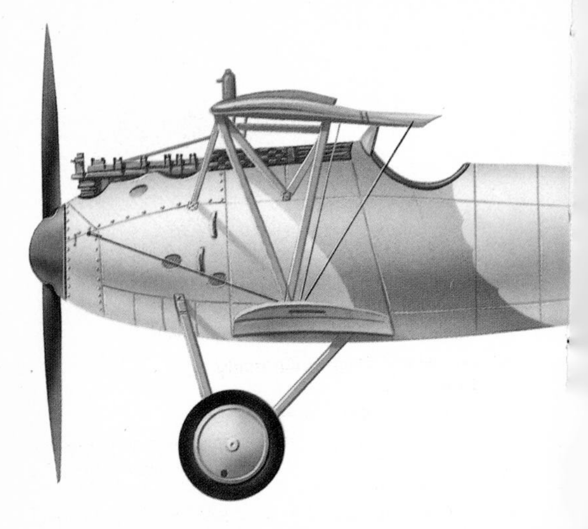

The Albatros D.Va superseded the D.III and was a beautifully streamlined fighter. By March 1918 it was not superior to the Allied fighting scouts because of weakness in the lower wing. Total production was probably in excess of 3,000 and 1,512 D.Vs and D.Vas served in the Western Front.

Country of Origin: Germany
Type: fighter
Accommodation: one
Armament: two 7.92mm Spandau machine-guns
Power Plant: one 180 or 200hp Mercedes D.IIIa water-cooled in-line
Performance: maximum speed 187km/h (116mph); service ceiling 6,250m (20,505ft); endurance 2 hours
Weight: take-off 937kg (2,066lb)
Dimensions: span 9.05m (29ft 8.25in); length 7.32m (24ft .5in); height 2.7m (8ft 10.25in); wing area 21.2m^2 (229sq ft)
Year entered service: 1917

Bristol F.2b

The Bristol F.2b re-equipped squadrons in France from the summer of 1917 and helped establish Allied air superiority by spring 1918. Some 5,250 were ordered and they remained in service after the war.

Country of Origin: Great Britain
Type: reconnaissance; fighter
Accommodation: two
Armament: one 8mm (.303in) Vickers machine-gun; one/two 8mm (.303in) Lewis machine-guns in rear cockpit; up to twelve 11kg (25lb) Cooper bombs under lower wings.
Power Plant: 275hp Rolls-Royce Falcon III
Performance: maximum speed 198km/h (123mph) at 1,220m (4,000ft); service ceiling 6,550m (21,500ft); endurance 3 hours
Weights: empty 770kg (1,700lb); loaded 1,200kg (2,650lb)
Dimensions: span 12m (39ft 3in); length 7.9m (25ft 10in); height 2.8m (9ft 4in)
Year entered service: 1917

Caproni Ca 5

The Caproni Ca 5 was the last wartime bomber produced by Italy. It marked a return to the biplane configuration and operated with Italian night-bomber squadrons in France during the last months of World War I. Total production reached 255 in Italy but some were built in France and America.

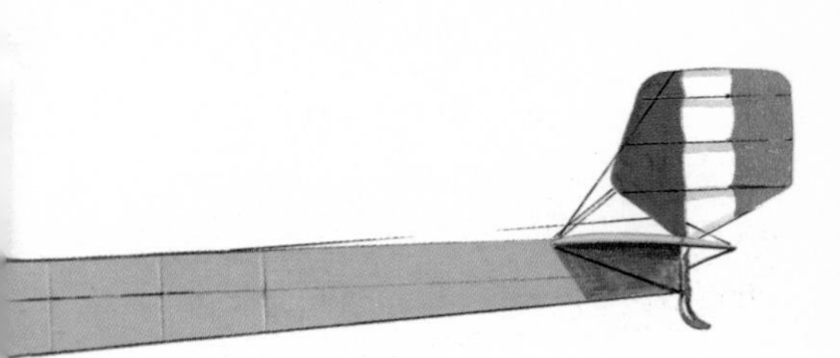

Country of Origin: Italy
Type: night bomber
Accommodation: three
Armament: two machine-guns; 540kg (1,190lb) bomb load
Power Plant: three 300hp Fiat A-12bis water-cooled in-line
Performance: maximum speed 152km/h (94mph) at sea level; service ceiling 4,500m (14,764ft); endurance 4 hours
Weight: take off 5,300kg (11,685lb)
Dimensions: span 23.4m (76ft 9.25in); length 12.62m (41ft 4.88in); wing area 150m^2 (1,614.6sq ft)
Year entered service: 1918

De Havilland D.H.9a

The D.H.9 replaced the D.H.4 and was planned for mass production. It moved the pilot behind the fuel tank close to the observer, however it had a low ceiling and poor manoeuvrability. Some 3,204 were built before it was replaced by the D.H.9a.

Country of Origin: Great Britain
Type: day bomber
Accommodation: two
Armament: one 8mm (.303in) Vickers machine-gun forward; one/two 8mm (.303in) Lewis machine-guns in rear cockpit; up to 227kg (500lb) of bombs on underwing racks
Power Plant: one 540hp Bristol Jupiter
Performance: maximum speed 179km/h (112mph) at 3,050m (10,000ft); service ceiling 5,334m (17,500ft); endurance 4½ hours
Weights: 999kg (2,203lb); loaded 1,664kg (3,669lb)
Dimensions: span 12.9m (42ft 4.63in); length 9.3m (30ft 6in); height 3.4m (11ft 2in)
Year entered service: 1918

Fokker D.VII

The Fokker D.VII was one of the great combat aircraft. It won the German standard fighter competition in January/February 1918. The first examples were sent to von Richthofen's JG1 in April 1918. Some 1,000 are thought to have been built.

Country of origin: Germany
Type: fighter
Accommodation: two fixed 7.92mm Spandau machine-guns firing forward
Power Plant: one 160hp Mercedes D.III or one 185hp BMWIII
Performance: maximum speed 189km/h (117mph) at 1,000m (3,280ft); service ceiling 6,000m (19,685ft); endurance 1½ hours
Weights: empty 686kg (1,513lb); take-off 900kg (1,984lb)
Dimensions: span 8.9m (29ft 3.5in); length 6.9m (22ft 9.75in); height 2.75m (9ft .25in); wing area $20.5m^2$ (221.4sq ft)
Year entered service: 1918

Fokker D.VIII

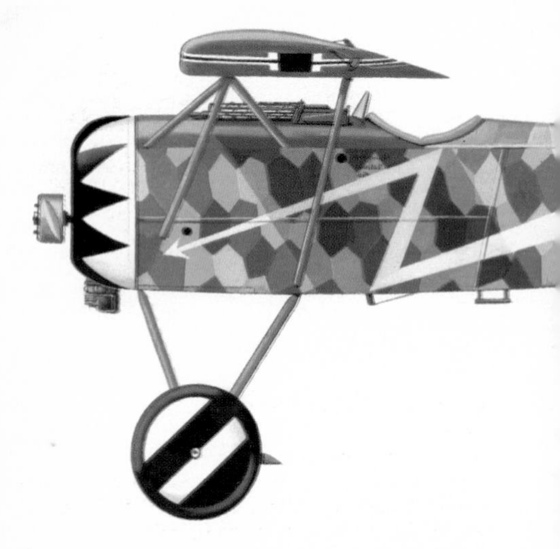

The Fokker D.VIII appeared too late to get into full service with front-line Jastas. It was a more manoeuvrable plane than the D.VII and some 400 were probably built

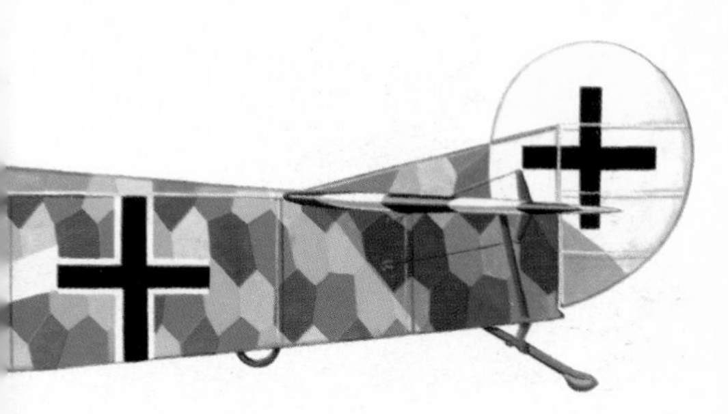

Country of Origin: Germany
Type: fighter
Accommodation: one
Armament: two 7.92mm Spandau machine-guns
Power Plant: one 110hp Oberursel UR.11 rotary
Performance: maximum speed 185km/h (115mph) at sea level; service ceiling 6,300m (20,669ft); endurance 1½ hours
Weight: take off 562kg (1,238lb)
Dimensions: span 8.4m (27ft 6.75in); length 5.86m (19ft 2.75in); wing area $10.7m^2$ (115.2sq ft)
Year entered service: 1918

Fokker Dr.I

The Fokker Dr.I triplane was built in response to the appearance of the Sopwith Triplane. Manfred von Richthofen was flying his Dr.I when he was killed on 21 April 1918. It acquired a bad reputation after a series of crashes, which were caused by faulty workmanship in the wing construction.

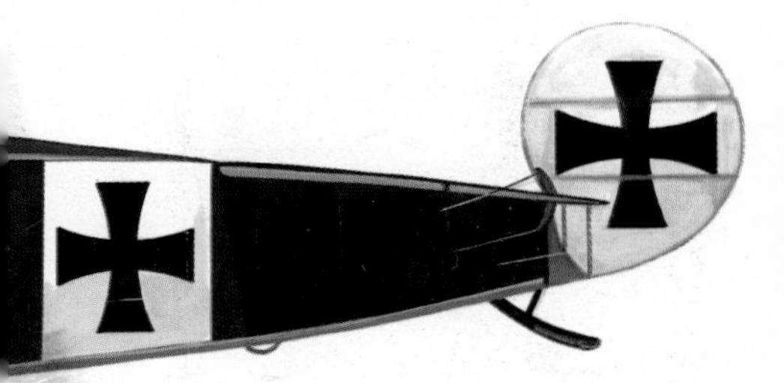

Country of Origin: Germany
Type: fighter
Accommodation: one
Armament: two 7.92mm Spandau machine-guns
Power Plant: one 110hp Oberursel UR.11 rotary
Performance: maximum speed 165km/h (102.5mph) at 4,000m (13,123ft); service ceiling 6,100m (20,013ft); endurance 1½ hours
Weight: take off 585kg (1,290lb)
Dimensions: span 7.2m (23ft 7.88in); length 5.77m (18ft 11.16in); wing area 18.7m² (200.9sq ft)
Year entered service: 1917

Fokker E.III

The Fokker E.III was the most numerous of the Fokker E types. Some 266 are thought to have been built and it made a significant contribution to the success of the 'Fokker Scourge'.

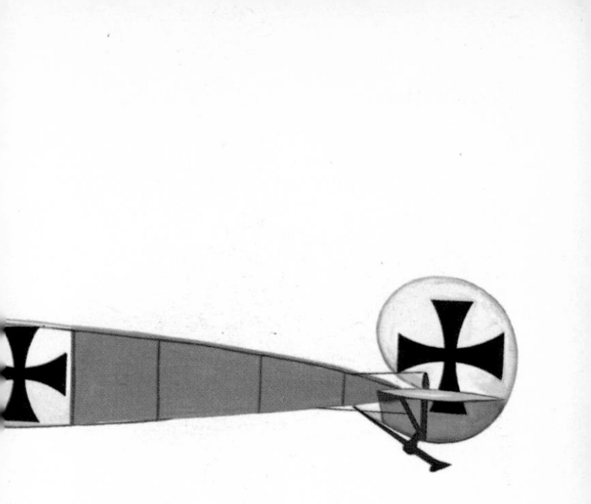

Country of Origin: Germany
Type: fighter
Accommodation: one
Armament: one 7.92mm Spandau machine-gun
Power Plant: one 100hp Oberursel U.1 rotary
Performance: maximum speed 133.6km/h (83mph) at 1,981m (6,500ft); service ceiling 3,500m (11,500ft); endurance $2\frac{3}{4}$ hours
Weight: take off 635kg (1,400lb)
Dimensions: span 9.52m (31ft 2.75in); length 7.3m (23ft 11.33in); wing area 16m^2 (172.2sq ft)
Year entered service: 1915

Gotha G.V

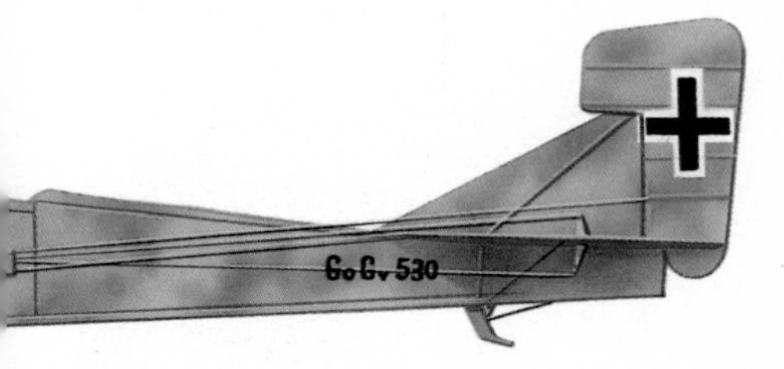

Country of Origin: Germany
Type: bomber
Accommodation: two
Armament: three machine-guns; 500kg (1,100lb) bomb load
Power Plant: two 260hp Mercedes D.IVa water-cooled in-line
Performance: maximum speed 140km/h (87mph) at sea level; service ceiling 6,250m (20,500ft); range 840km (520 miles)
Weight: take off 3,975kg (8,763lb)
Dimensions: span 23.7m (77ft 9.16in); length 12.35m (40ft 6.25in); wing area 89.5m^{2}2 (963.4sq ft)
Year entered service: 1917

Halberstadt C.V

The Halberstadt C family of reconnaissance planes made an important contribution to the strength of the Central Powers. The C.V appeared in early 1918. Its reconnaissance camera was aimed downward through a trap in the floor of the rear cockpit.

Country of Origin: Germany
Type: bomber/long-range photographic reconnaissance
Accommodation: two
Armament: Parabellum machine-guns and 7.92mm Spandau machine-gun
Power Plant: one 220hp Benz Bz.IV water-cooled in-line
Performance: maximum speed 170km/h (106mph); service ceiling 5,000m (16,400ft); endurance 3½ hours
Weight: take off 1,365kg (3,009lb)
Dimensions: span 13.6m (44ft 8.3in); length 6.9m (22ft 8.5in); height 3.4m (11ft .25in); wing area 43m² (462.8sq ft)
Year entered service: 1918

Halberstadt D.II

The Halberstadt D.II looked frail but was very strong and manoeuvrable. It served as an escort for two-seat reconnaissance aircraft before joining front-line *Jagdstafflen*. By 1918 it was transferred from the Western Front to Macedonia and Palestine.

Country of Origin: Germany
Type: fighter
Accommodation: two
Armament: one 7.92mm Spandau machine-gun
Power Plant: one 120hp Mercedes D.II water-cooled in-line
Performance: maximum speed 145km/h (90mph) at sea level; service ceiling, approx 13,000ft; endurance, approx 1½ hours
Weight: take off 730kg (1,609lb)
Dimensions: span 8.8m (28ft 10.5in); length 7.3in (23ft 11.38in); wing area, approx 22.8m² (245sq ft)
Year entered service: 1916

Handley Page O/400

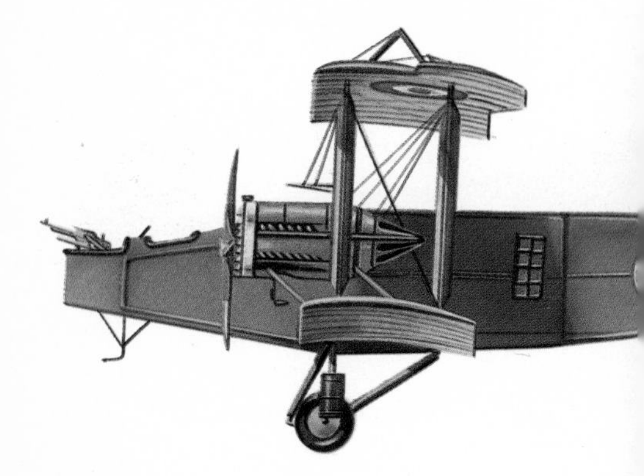

Country of Origin: Great Britain
Type: night bomber
Accommodation: four
Armament: two 8mm (.303in) Lewis machine-guns in rear cockpit; one or two 8mm (.303in) Lewis machine-guns in nose cockpit; up to sixteen 51kg (112lb) bombs, or one 748kg (1,650lb) SN bomb; one 8mm (.303in) Lewis machine-gun firing through floor of rear cockpit

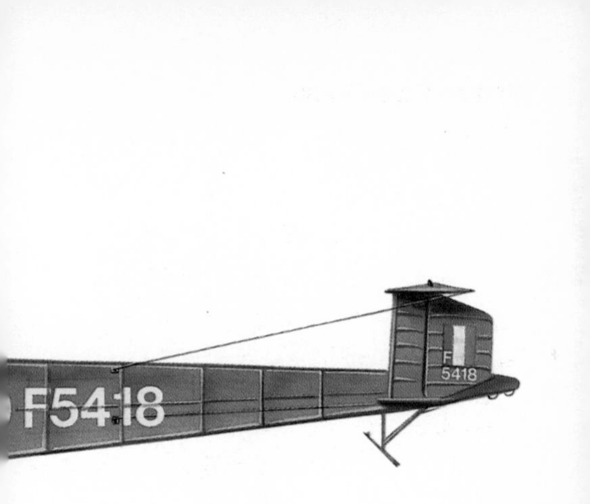

Power Plant: two 360hp Rolls-Royce Eagle VIII
Performance: maximum speed 157km/h (97.5mph) at ground level; service ceiling 2,590m (8,500ft); endurance 8 hours
Weights: empty 3,857kg (8,502lb); loaded 6,060kg (13,360lb)
Dimensions: upper span 30.5m (100ft); lower span 21.3m (70ft); length 19.2m (62ft 10.25in); height 6.7m (22ft)
Year entered service: 1917

Handley Page V/1500

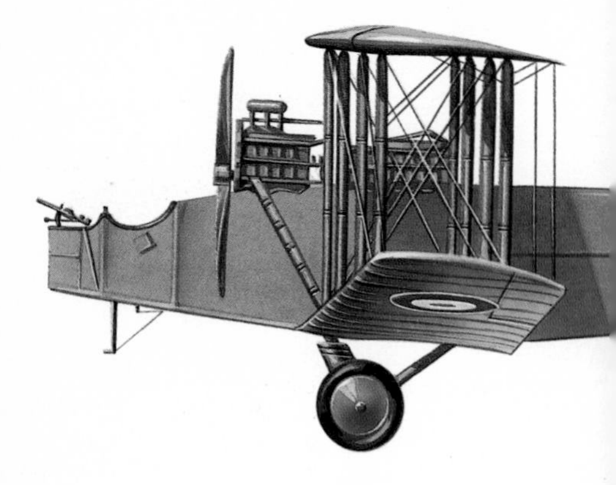

The Handley Page V/1500 was the first British four-engined bomber to go into production and was the largest British aircraft built during World War I. Only three were delivered before the Armistice so only 36 were built in all.

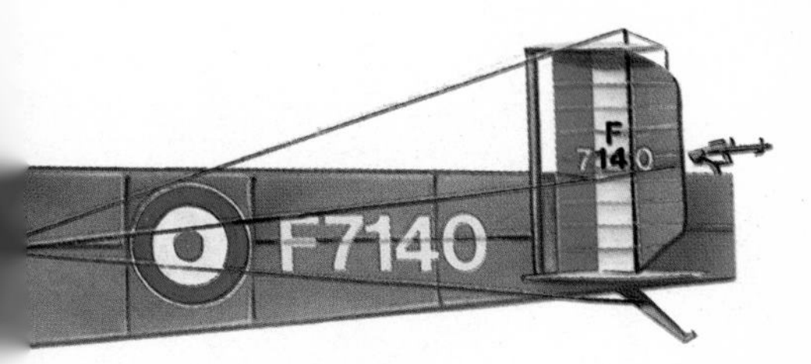

Country of Origin: Great Britain
Type: strategic night bomber
Accommodation: six or seven
Armament: single 8mm (.303in) Lewis machine-gun in nose, dorsal, ventral and tail cockpits; bomb load up to thirty 113kg (250lb) bombs
Power Plant: four 375hp Rolls-Royce Eagle VIII
Performance: maximum speed 146km/h (90.5mph) at 1,980m (6,500ft); service ceiling 3,350m (11,000ft); endurance 14 hours
Weights: empty 7,984kg (17,602lb); loaded 10,920kg (24,080lb)
Dimensions: span 38.4m (126ft); length 18.9m (62ft); height 7m (23ft)
Year entered service: 1918

Hansa-Brandenburg W.12

Designed by Ernst Heinkel, the 146 Hansa-Brandenburg W.12s were active along the Flanders coast and over the North Sea.

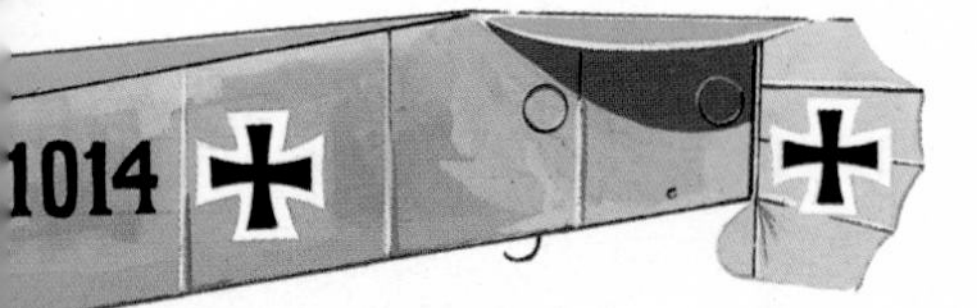

Country of Origin: Germany
Type: seaplane
Accommodation: two
Armament: Parabellum gun
Power Plant: one 150hp Benz Bz.III water-cooled in-line
Performance: maximum speed 160km/h (99mph) at sea level; service ceiling 5,000m (16,400ft); endurance 3½ hours
Weight: take-off 1,465kg (3,230lb)
Dimensions: span 11.2m (36ft 9in); length 9.65m (31ft 7.8in); wing area 36.2m^2 (389.7ft)
Year entered service: 1917

Hansa-Brandenburg W.29

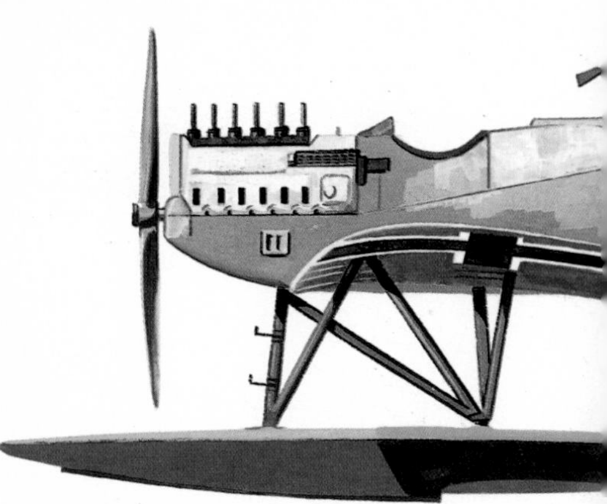

The Hansa-Brandenburg W.29 was a monoplane version of the W.12. It served alongside the W.12 and W.19 and had a good combat record. Some 78 were delivered before the Armistice.

Country of Origin: Germany
Type: seaplane
Accommodation: two

Armament: variable twin guns and Parabellum gun
Power Plant: one 150hp Benz B.III water-cooled in-line
Performance: maximum speed 170km/h (106mph) at sea level; service ceiling over 3,000m (9,843ft); endurance 4 hours
Weight: take off 1,420kg (3,131lb)
Dimensions: span 13.5m (44ft 3.5in); length 9.3m (30ft 6.25in); wing area 31.6m^2 (340sq ft)
Year entered service: 1918

Martinsyde G.102 Elephant

Country of Origin: Great Britain
Type: fighter (originally); bomber
Accommodation: one
Armament: one 8mm (.303in) Lewis machine-gun on upper-wing mounting; one 8mm (.303in) Lewis machine-gun on cockpit mounting; up to 113kg (250lb) of bombs under wings and/or fuselage
Power Plant: one 160hp Beardmore

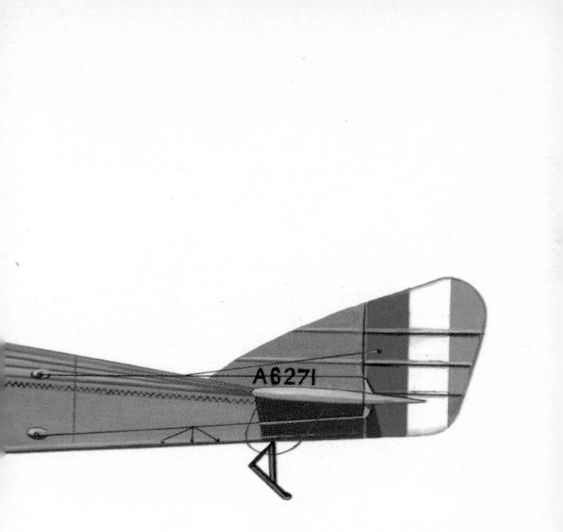

Performance: maximum speed 167km/h (104mph) at 3,048m (2,000ft); service ceiling 4,877m (16,000ft); endurance 4½ hours
Weights: empty 813kg (1,793lb); loaded 1,075kg (2,370lb)
Dimensions: span 11.6m (38ft); length 8.2m (27ft); height 2.9m (9ft 8in); wing area 38.1m^2 (410sq ft)
Year entered service: 1917

Morane-Saulnier Type L

The Morane-Saulnier Type L was a parasol monoplane, with its wing mounted above the fuselage. In spite of being a two-seater it was faster than most German machines of the time.

Country of Origin: France
Type: parasol monoplane fighter
Accommodation: two

Armament: variable; Hotchkiss machine-gun and propeller blade deflectors; could carry six 10kg (25lb) bombs
Power Plant: one 80hp Gnome rotary
Performance: maximum speed 115km/h (71.5mph) at 2,000m (6,560ft); service ceiling 4,000m (13,123ft); range 450km (280 miles)
Weight: take off 680kg (1,499lb)
Dimensions: span 11.2m (36ft 9in); length 6.88m (22ft 6.75in); wing area 18.3m^2 (197sq ft)
Year entered service: 1913

Morane-Saulnier Type N

The Morane-Saulnier Type N was built in small numbers (49) but saw plenty of service in the first year of the war. It had good speed and could climb to 3,000m (9,840ft) in 12 minutes.

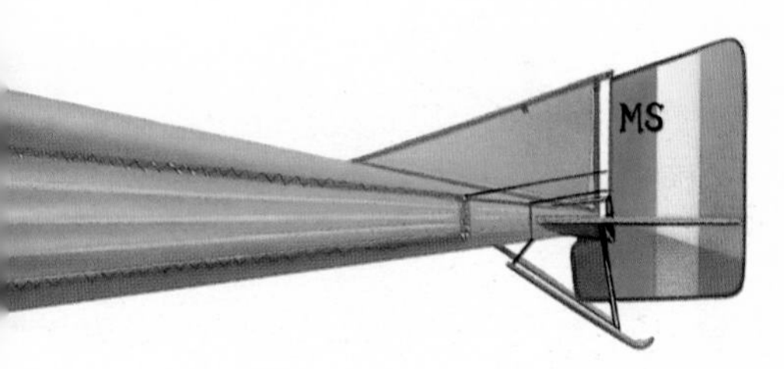

Country of Origin: France
Type: fighter
Accommodation: one
Armament: one 8mm (.303in) Lewis machine-gun
Power Plant: one 80hp Le Rhône 9C rotary
Performance: maximum speed 144km/h (89.5mph) at sea level; service ceiling 4,000m (13,123ft); endurance $1\frac{1}{2}$ hours
Weight: take off 444kg (979lb)
Dimensions: span 8.15m (26ft 8.75in); length 5.83m (19ft 1.5in); wing area 11m^2 (118.4sq ft)
Year entered service: 1914

Nieuport XII

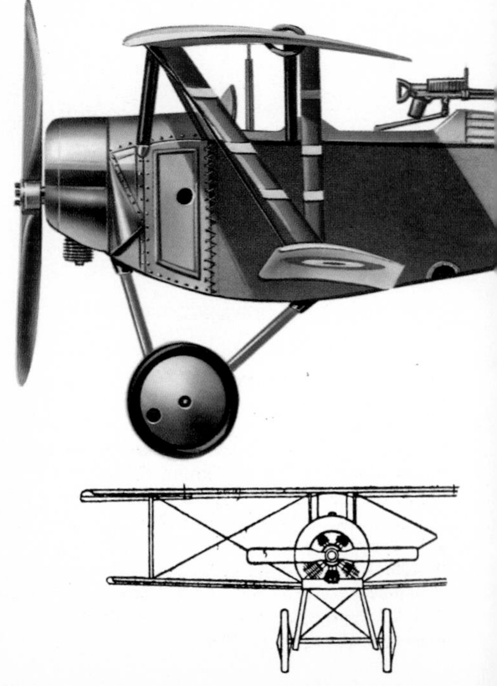

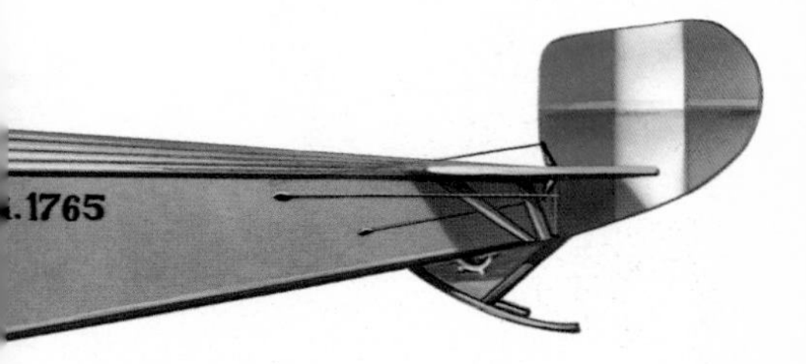

Country of Origin: France
Type: observation
Accommodation: two
Armament: one 8mm (.303in) Lewis machine-gun
Power Plant: one 130hp Clerget 9B rotary
Performance: maximum speed 155km/h (96.2mph) at sea level; service ceiling 4,700m (15,420ft); endurance, approx $2\frac{3}{4}$ hours
Weight: take off 920kg (2,028lb)
Dimensions: span 9.03m (29ft 7.5in); length 7.3m (23ft 11.88in); wing area $22m^2$ (236.8sq ft)
Year entered service: 1915

Nieuport Scout XVII

The Nieuport XVII ended the 'Fokker Scourge' and was one of the most successful aircraft of World War I. Guynemer, Bishop, Ball and Boyau were among the aces who flew this Nieuport.

Country of Origin: France
Type: fighter
Accommodation: one
Armament: one 8mm (.303in) Lewis machine-gun mounted above upper wing
Power Plant: one 110hp Le Rhone 9Ja
Performance: maximum speed 166km/h (103mph) at ground level; service ceiling 4,799m (15,744ft); endurance 1¾ hours
Weights: empty 374kg (825lb); loaded 559kg (1,232lb)
Dimensions: upper span 8.1m (26ft 9in); lower span 7.8m (25ft 7in); length 5.8m (19ft); height 2.4m (7ft 10in)
Year entered service 1915

Nieuport Scout XXVII

The Nieuport XXVII was the peak of development of the basic 'V-strutters'. Some 120 were built.

Country of Origin: France
Type: fighter
Accommodation: one
Armament: one 8mm (.303in) Vickers machine-gun fixed on forward fuselage
Power Plant: one 120hp Le Rhone 9Jb
Performance: 172km/h (107mph) at ground level; service ceiling 6,858m (22,500ft); endurance 2¼ hours
Weights: 355kg (783lb); loaded 535kg (1,179lb)
Dimensions: upper span 8.2m (26ft 11in); lower span 7.8m (25ft 8in); length 5.9m (19ft 3in); height 2.4m (7ft 10in)
Year entered service: 1918

Pfalz D.III

The Pfalz D.III was an extremely serviceable fighter which saw a great deal of action with front-line *Jastas*. It was not as fast as the Albatros D.Va nor did it have the altitude performance of the Fokker D.VII but it made a tremendous contribution to German air superiority in early 1918.

Country of Origin: Germany
Type: fighter
Accommodation: one
Armament: twin 7.92mm Spandau guns
Power Plant: one 160hp Mercedes D.III water-cooled in-line
Performance: maximum speed 165km/h (102.5mph) at 3,000m (9,843ft); service ceiling 5,200m (17,060ft); endurance $2\frac{1}{2}$ hours
Weight: take off 932kg (2,055lb)
Dimensions: span 9.4m (30ft 10.16in); length 6.95m (22ft 9.66in); wing area 22.17m^2 ((238.6sq ft)
Year entered service: 1917

Royal Aircraft Factory B.E.12

Royal Aircraft Factory B.E. 12
The B.E.12 was based on the B.E.2c. After only a few weeks it was withdrawn from service as a fighter because it was vulnerable to air attack but was reinstated as a light bomber.

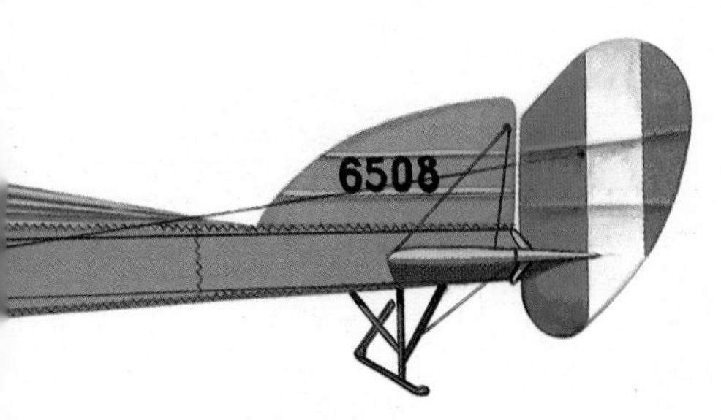

Country of Origin: Great Britain
Type: fighter/light bomber
Accommodation: one
Armament: sixteen 7.25kg (16lb) or two 51kg (112lb) bombs
Power Plant: one 150hp RAF4a air-cooled Vee- type
Performance: maximum speed 164km/h (102mph); service ceiling 3,810m (12,500ft); endurance 3 hours
Weight: take off 1,067kg (2,352lb)
Dimensions: span 11.3m (37ft); length 8.3m (27ft 3in); height 3.39m (11ft 1½in)
Year entered service: 1916

Royal Aircraft Factory F.E.2b

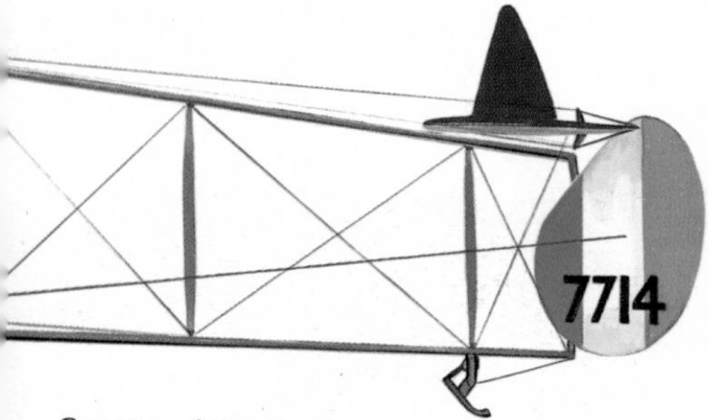

Country of Origin: Great Britain
Type: fighter, reconnaissance, bomber
Accommodation: two
Armament: one or two 8mm (.303in) Lewis machine-guns in forward cockpit; up to 136kg (300lb) of bombs on external racks
Power Plant: 120/160hp Beardmores
Performance: maximum speed 147km/h (91.5mph) at ground level; service ceiling 3,353m (11,000ft); endurance 3 hours
Weights: empty 935kg (2,061lb); loaded 1,378kg (3,037lb)
Dimensions: 14.5m (47ft 9in); length 9.8m (32ft 3in); height 3.8m (12ft 5in)
Year entered service: 1914

Royal Aircraft Factory S.E.5

The S.E.5 was faster and stronger than the Spads or Nieuports, even if it was not as manoeuvrable as the latter. With the Camel it enabled the Allies to regain air superiority in 1918. By the end of the war some 2,700 were in service with the R.F.C.

Country of Origin: Great Britain
Type: fighter
Accommodation: one

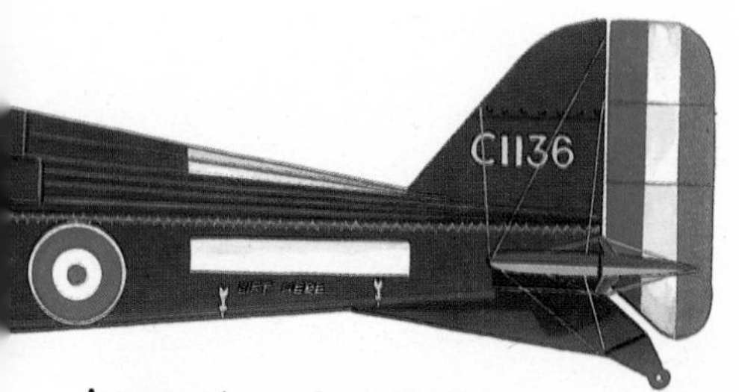

Armament: one 8mm (.303in) Lewis machine-gun on upper wing mounting; one fixed 8mm (.303in) Vickers machine-gun on forward fuselage; four 11.3kg (25lb) Cooper bombs racked under fuselage
Power Plant: one 150hp Hispano-Suiza SA water-cooled
Performance: maximum speed 207km/h (128.5mph) at ground level; service ceiling 5,790m (19,000ft); endurance 2½ hours
Weights: empty 581kg (1,280lb); loaded 829kg (1,827lb)
Dimensions: span 8.5m (27ft 11in); length 6.4m ((20ft 11in); height 2.9m (9ft 6in)
Year entered service: 1917

Siemens-Schuckert D.III

The Siemens-Schuckert D.III showed great promise with good speed, climb and manoeuvrability. Some 80 were built before it was superseded by the D.IV, which was described in October 1918 as 'superior by far to all single-seaters in use on the Front today'.

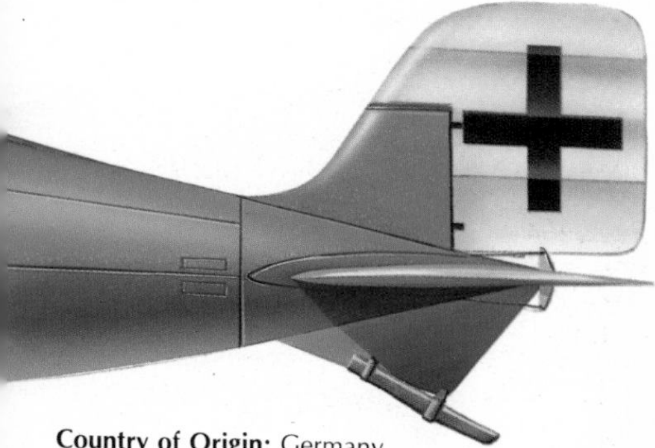

Country of Origin: Germany
Type: fighter
Accommodation: one
Armament: twin 7.92mm Spandau machine-guns
Power Plant: one 160hp Siemens-Halske Sh.III rotary
Performance: maximum speed 180km/h (111.8mph) at sea level; service ceiling 8,100m (26,575ft); endurance 2 hours
Weight: take off 725kg (1,598lb)
Dimensions: span 8.43m (27ft 7.88in); length 5.7m (18ft 8.38in); wing area 18.8m^2 (202.6sq ft)
Year entered service: 1918

Siemens-Schuckert R.I

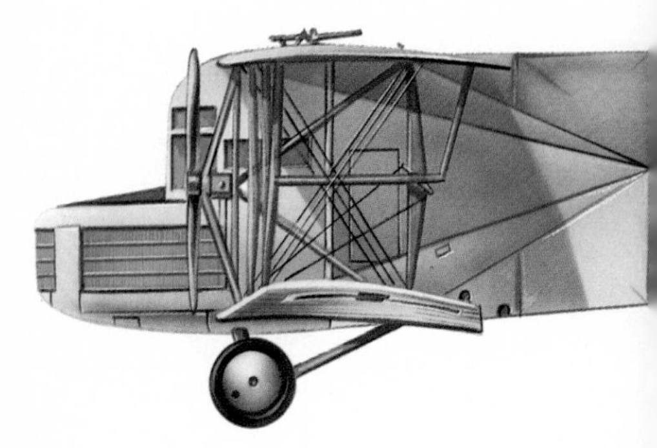

The Siemens-Schukert R-types were some of the largest German aircraft of World War I. The R.IV to R.VII were employed on operational duties but the R.I, R.II and R.III were used as trainers in Germany.

Country of Origin: Germany
Type: bomber
Accommodation: three
Armament: one 7.92mm Spandau machine-gun
Power Plant: three 150hp Benz Bz.III water-cooled in-line
Performance: maximum speed 110km/h (68.4mph); sevice ceiling 3,800m (12,467ft); endurance 4 hours
Weight: take off 5,200kg (11,464lb)
Dimensions: span 28m (91ft 10.25in); length 17.5m (57ft 5in); wing area 138m^2 (1,485.4sq ft)
Year entered service: 1916

Sikorsky Ilya Mouromets

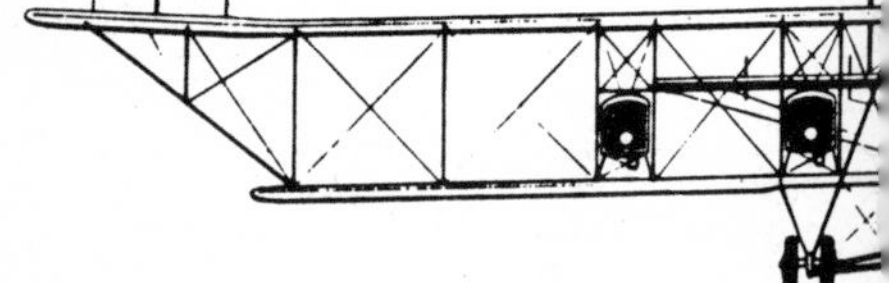

Country of Origin: Russia
Type: bomber
Accommodation: five
Armament: three/four machine-guns; 450–700kg (992–1,543lb) bomb load
Power Plant: four 150hp Sunbeam water-cooled Vee-type

Performance: maximum speed 110km/h (68.4mph) at 2,000m (6,562ft); service ceiling 2,900m (9,514ft); endurance 4 hours
Weight: take off 4,600kg (10,141lb)
Dimensions: span 29.8m (97ft 9.13in); length 17.5m (57ft 5in); wing area 125m^2 (1,345.5sq ft)
Year entered service: 1914

Sopwith $1\frac{1}{2}$ Strutter

The Sopwith $1\frac{1}{2}$-Strutter was the first aircraft to be designed with a synchronised gun (apart from Sikorsky Ilya Mouromets) and the first to equip a strategic bombing force. Some 1,513 were built in Great Britain and in 1917 it was mainly used for coastal patrols and as a night fighter.

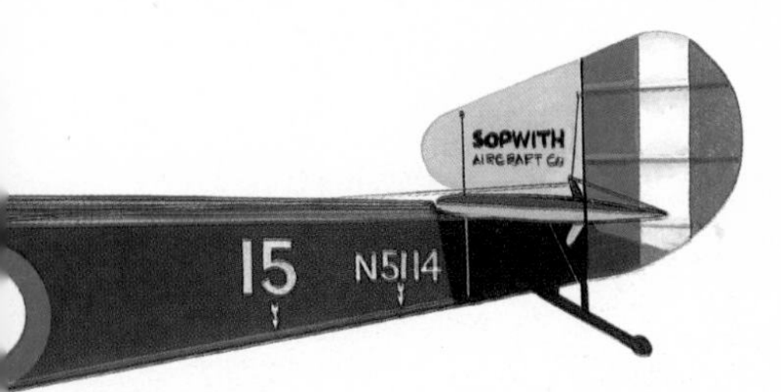

Country of Origin: Great Britain
Type: day bomber
Accommodation: two
Armament: one fixed 8mm (.303in) Vickers machine-gun forward; one 8mm (.303in) Lewis machine-gun in rear cockpit; up to 104kg (230lb) of bombs
Power Plant: one 110hp Clerget 9z
Performance: maximum speed 171km/h (106mph) at ground level; service ceiling 4,570m (15,000ft); endurance 4½ hours
Weights: empty 571kg (1,259lb); loaded 975kg (2,149lb)
Dimensions: span 10.2m (33ft 6in); length 7.7m (25ft 3in); height 3.1m (10ft 3in)
Year entered service: 1916

Sopwith Camel F.1

The Sopwith Camel was probably the most famous World War I aircraft. It was a brilliant dogfighter and total production reached 5,490. The Camel's most famous victim was Manfred von Richthofen.

Country of Origin: Great Britain
Type: fighter
Accommodation: one
Armament: two fixed 8mm (.303in) Vickers machine-guns forward; four 11kg (25lb) Cooper bombs under fuselage if required
Power Plant: one 130hp Clerget
Performance: maximum speed 182km/h (113mph) at 3,050m (10,000ft); service ceiling 5,790m (19,000ft); range 400km (250 miles)
Weights: empty 421kg (929lb); loaded 659kg (1,453lb)
Dimensions: span 8.5m (28ft); length 5.7m (18ft 9in); height 12.1m (8ft 6in); wing area 21.5m^2 (231sq ft)
Year entered service: 1917

Sopwith Dolphin Mk.1

The Sopwith Dolphin was designed to give the pilot a better view with increased armament. It was a tough fighter and some 1,532 were delivered to front-line units.

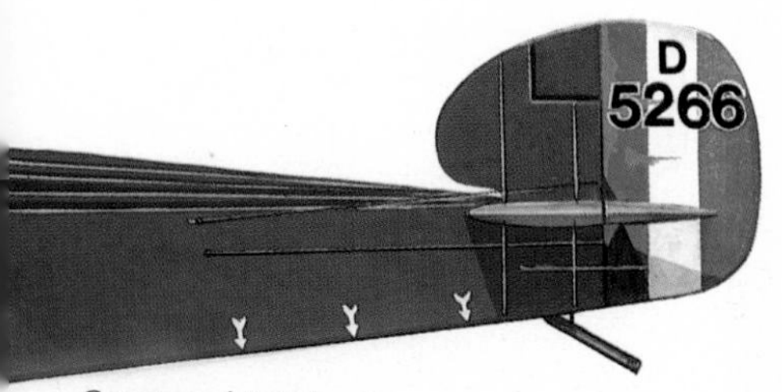

Country of Origin: Great Britain
Type: fighter
Accommodation: one
Armament: two fixed 8mm (.303in) Vickers machine-guns forward, plus if required one or two 8mm (.303in) Lewis machine-guns on upper wing mountings; four 11kg (25lb) Cooper bombs if required
Power Plant: one 200hp Hispano-Suiza
Performance: maximum speed 206km/h (128mph) at 3,050m (10,000ft); service ceiling 6,400m (21,000ft); endurance 2 hours
Weights: empty 651kg (1,436lb); loaded 867kg (1,911lb)
Dimensions: span 9.9m (32ft 6in); length 6.8m (22ft 3in); height 2.6m (8ft 6in)
Year entered service: 1917

Sopwith Pup

The Sopwith Pup was a small, simple and reliable aircraft. It was a delight to fly but it was underpowered and was withdrawn from front-line service in 1918. Some 1,840 were probably built.

Country of Origin: Great Britain
Type: fighter
Accommodation: one

Armament: one 8mm (.303in) Vickers machine-gun forward (standard), or one 8mm (.303in) Lewis machine-gun firing through or above centre section of upper wing; four 9kg (20lb) Hales bombs could be carried under fuselage
Power Plant: one 80hp Le Rhône
Performance: maximum speed 179km/h (111.5mph) at ground level; service ceiling 5,330m (17,500ft); endurance 3 hours
Weights: empty 357kg (787lb); loaded 556kg (1,225lb)
Dimensions: span 8.1m (26ft 6in); length 5.9m (19ft 3.75in); height 2.9m (9ft 5in); wing area 23.6m^2 (254sq ft)
Year entered service: 1916

Sopwith Snipe

The Sopwith Snipe was designed as a successor to the Camel and gave the pilot a better view. It was faster, more manoeuvrable and more reliable than its predecessor. Deliveries to France only started eight weeks before the Armistice but it had an excellent service record in a short period of time.

Country of Origin: Great Britain
Type: fighter
Accommodation: one
Armament: two fixed 8mm (.303in) Vickers machine-guns forward; four 11kg (25lb) Cooper bombs under fuselage if required
Power Plant: one 230hp Bentley B.R.2 rotary
Performance: maximum speed 195km/h (121mph) at 3,050m (10,000ft); service ceiling 5,940m (19,500ft); endurance 3 hours
Weights: empty 595kg (1,312lb); loaded 916kg (2,020lb)
Dimensions: upper wing span with balanced ailerons 9.4m (31ft 1in); length 6m (19ft 10in); height 2.9m (9ft 6in); wing area 23.8m^2 (256sq ft)
Year entered service: 1918

Sopwith Triplane

The Sopwith Triplane greatly perturbed the Germans when it first appeared. It had an excellent climb rate and combat record and saw service with the R.N.A.S. It was replaced by the Camel.

Country of Origin: Great Britain

Type: fighter

Accommodation: one

Armament: one fixed 8mm (.303in) Vickers machine-gun forward

Power Plant: one 130hp Clerget

Performance: maximum speed 188km/h (117mph) at 1,520m (5,000ft); service ceiling 6,250m (20,500ft); endurance 2¾ hours

Weights: empty 499kg (1,101lb); loaded 699kg (1,541lb)

Dimensions: span 8.1m (26ft 6in); length 5.7m (18ft 10in); height 3.2m (10ft 6in)

Year entered service: 1916

Spad VII

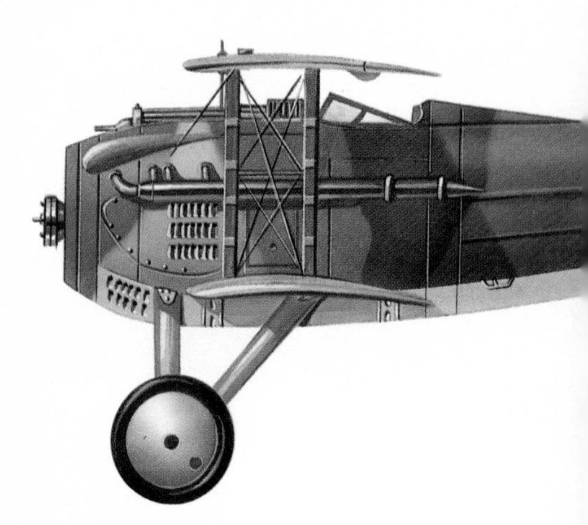

Spad produced some of the best combat aircraft of World War I. The Spad VII was a strong, stable gun platform with excellent turn of speed and good climb to 3,660m (12,000ft). It saw widespread service with every Allied air service. Some 6,000 were probably built.

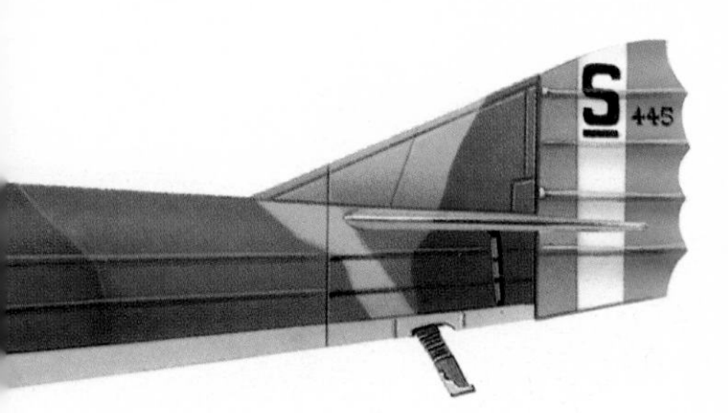

Country of Origin: France
Type: fighter
Accommodation: one
Armament: one fixed 8mm (.303in) Vickers machine-gun forward
Power Plant: one 150hp Hispano-Suiza
Performance: maximum speed 192km/h (119mph) at 1,980m (6,500ft); service ceiling 5,340m (17,500ft); endurance $2\frac{1}{4}$ hours
Weight: loaded 740kg (1,632lb)
Dimensions: span 7.8m (25ft 8in); length 6.1m (20ft 3.5in); height 2.1m (7ft)
Year entered service: 1916

Spad XIII

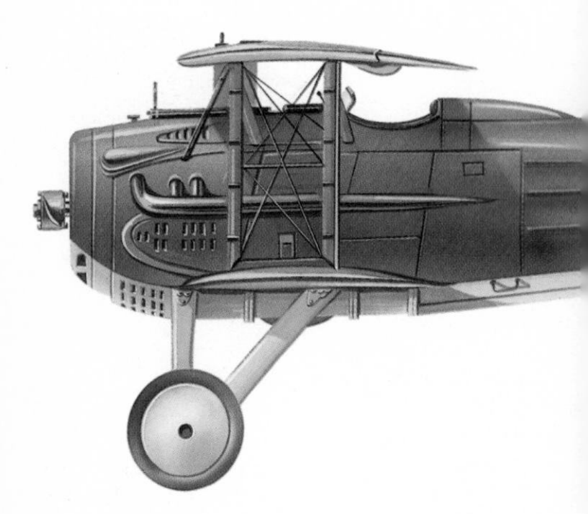

The Spad XIII replaced the VII and was built in greater numbers than any other Allied aircraft – some 8,472 were built. It was very fast, strong and reliable and equipped over 80 *escadrilles*. The ace Guynemer met his death while flying a Spad XIII.

Country of Origin: France
Type: fighter
Accommodation: one
Armament: two 8mm (.303in) Vickers machine-guns
Power Plant: one 235hp Hispano-Suiza 8 Be water-cooled Vee-type
Performance: maximum speed 215km/h (134mph) at 2,000m (6,562ft); service ceiling 6,650m (21,818ft); endurance 2 hours
Weight: take off 820kg (1,808lb)
Dimensions: span 8.08m (26ft 4.88in); length 6.22m (20ft 4.88in); wing area 21.2m^2 (227.7sq ft)
Year entered service: 1917

Aichi D3A1 'Val'

Country of Origin: Japan
Type: carrier-borne dive-bomber
Accommodation: two
Armament: two 7.7mm Type 97 machine-guns in engine cooling plus one flexible 7.7mm Type 92 machine-gun in rear cockpit, maximum bomb load 370kg (820lb)
Power Plant: one 1,000hp Mitsubishi Kinsei 43 or 1,080hp Kinsei 44
Performance: maximum speed 386km/h (240mph) at 3,000m (9,840ft); service ceiling 9,300m (30,050ft); range 1,472km (915 miles)
Weights: empty 2,408kg (5,309lb); loaded 3,650kg (8,047lb)
Dimensions: span 14.35m (47ft 2in); length 10.2m (33ft 5.5in); height 3.3m (10ft 11.25in); wing area 34.9m^2 (375.7sq ft)
Year entered service: 1939

Arado Ar 234B-2 Blitz

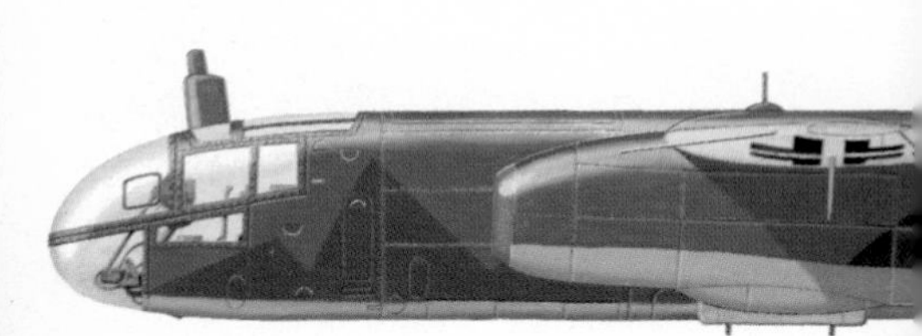

The first 234 took off from a three-wheel trolley on 15 June 1943 landing back on skids. However after extensive trials conventional landing gear was fitted before major production began. Main production of the 234B-2 was of a variant able to carry a heavy bomb load and these saw service at the Battle of the Bulge in 1944–45 and on attacks on the Remagen bridge across the Rhine.

Country of Origin: Germany
Type: twin-turbojet bomber aircraft

Accommodation: one
Armament: two 20mm MG151/20 cannon; maximum bomb load 1,500kg (3,308lb)
Power Plant: two 890kg (1,980lb) thrust Junkers Jumo 004B
Performance: maximum speed 735km/h (457mph) at 6,000m (19,685ft); service ceiling 10,000m (32,810ft); range 1,630km (1,013 miles)
Weights: empty 5,200kg (11,464lb); maximum 9,800kg (21,715lb)
Dimensions: span 14.4m (47ft 3.25in); length 12.6m (41ft 5.25in); height 4.3m (14ft 1.25in); wing area 26.4m^2 (284.2sq ft)
Year entered service: 1944

Armstrong Whitworth Whitley Mk V

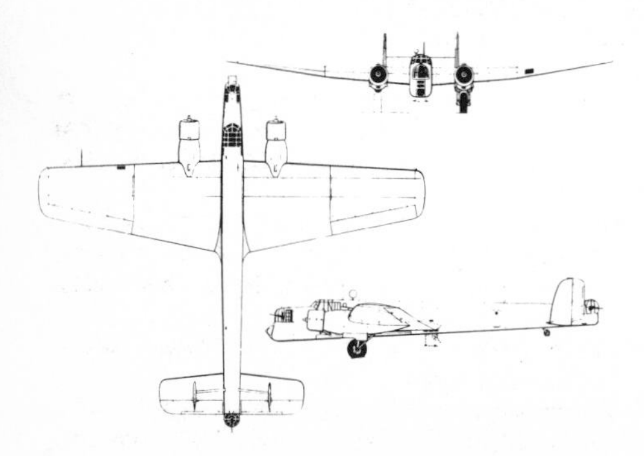

Armstrong Whitworth Whitley III

Country of Origin: Great Britain
Type: night bomber
Accommodation: five (normal)
Armament: one or two 8mm (.303in) Vickers machine-guns in nose turret; four 8mm (.303in) Browning machine-guns in tail turret; up to 3,175kg (7,000lb) bombs
Power Plant: two 1,145hp Rolls-Royce Merlin X
Performance: maximum speed 370km/h (230mph); service ceiling 7,925m (26,000ft); normal range 2,414km (1,500 miles)
Weights: empty 8,780kg (19,350lb); loaded 15,200kg (33,500lb)
Dimensions: span 25.6m (84ft); length 21.5m (70ft 6in); height 4.57m (15ft); wing area 105.6m^2 (1,137sq ft)
Year entered service: 1938

Avro Lancaster Mk 1

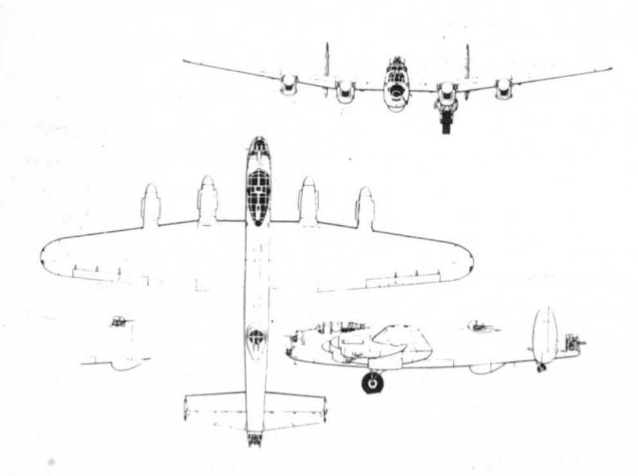

Avro Lancaster BIII

Country of Origin: Great Britain
Type: heavy bomber
Accommodation: seven or eight
Armament: two 8mm (.303in) Browning machine-guns each in nose and dorsal turret; four 8mm (.303in) Browning machine-guns in tail turret, or two 13mm (.5in) Browning machine-guns in tail rose turret; maximum bomb load 8,165kg (18,000lb)
Power Plant: four 1,280hp Rolls-Royce Merlin XX
Performance: maximum speed 462km/h (287mph); service ceiling 6,706m (22,000ft); maximum range 4,020km (2,500 miles)
Weights: empty 16,536kg (36,457lb); normal 29,480kg (65,000lb)
Dimensions: span 31m (102ft); length 21m (69ft 4in); height 6m (20ft 6in); wing area 120.5m^2 (1,297sq ft)
Year entered service: 1942

Bell P-39 Airacobra

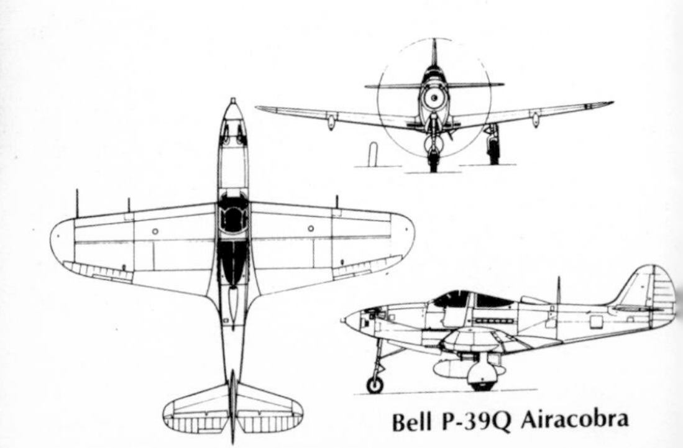

Bell P-39Q Airacobra

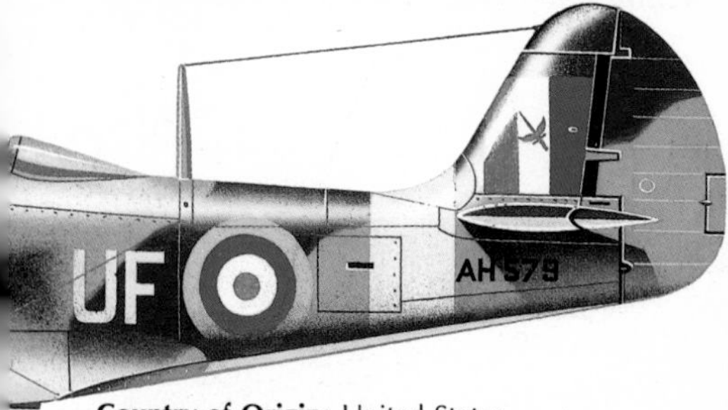

Country of Origin: United States
Type: interceptor fighter
Accommodation: one
Armament: one 37mm (1.45in) gun; four 13mm (.5in) guns; one 227kg (500lb) bomb
Power Plant: one 1,200hp Allison V-1710-35
Performance: maximum speed 539km/h (335mph) at 1,524m (5,000ft); service ceiling 10,670m (35,000ft); range 1,046km (650 miles)
Weights: empty 2,561kg (5,645lb); gross 3,765kg (8,300lb)
Dimensions: span 10.4m (34ft); length 9.2m (30ft 2in); height 3.7m (12ft 5in); wing area $19.8m^2$ (213sq ft)
Year entered service: 1941

Blohm und Voss Bv 138A-1

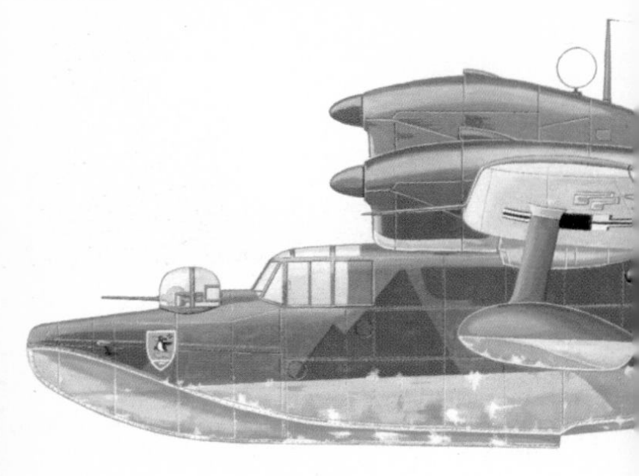

Country of Origin: Germany
Type: long-range maritime reconnaissance flying boat
Accommodation: five–six/ten
Armament: one 20mm MG 204 cannon in nose turret; one 7.9mm MG 15 machine-gun each in two open fuselage positions, one aft central engine nacelle and one at rear of hull; maximum bomb load 150kg (330lb)

Power Plant: three 600hp Junkers Jumo 205C-4
Performance: maximum speed 265km/h (165mph) at sea level; service ceiling 3,600m (11,810ft); maximum range 3,930km (2,442 miles)
Weights: empty 10,800kg (23,810lb); loaded 13,750kg (30,310lb)
Dimensions: span 26.9m (88ft 4in); length 19.8m (65ft 1.5in); height 5.9m (19ft 4.25in); wing area $112m^2$ (1,205sq ft)
Year entered service: 1940

Boeing B-17G Flying Fortress

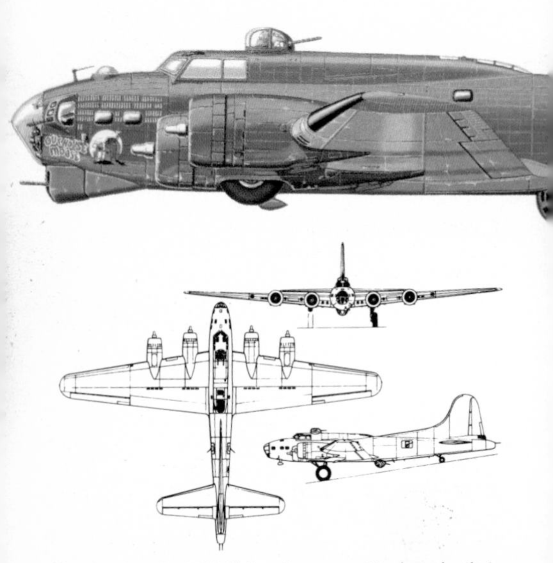

The Boeing B–17G Flying Fortress, 42 were built in 1940 and after Pearl Harbor were involved in the first US air offensive of WW2 on Japanese shipping.

Country of Origin: United States
Type: high-altitude bomber
Accommodation: ten
Armament: thirteen 13mm (.5in) machine-guns in chin, cheek, ventral, dorsal, waist and tail locations; up to 7,983kg (17,600lb) bombs
Power Plant: four 1,200hp Wright Cyclone R-1820-97
Performance: maximum speed 486km/h (302mph) at 7,620m (25,000ft); service ceiling 10,850m (35,600ft); maximum range 5,470km (3,400 miles)
Weights: empty 16,391kg (36,135lb); maximum 32,660kg (72,000lb)
Dimensions: span 31.6m (103ft 9in); length 22.8m (74ft 9in); height 5.8m (19ft 1in); wing area 131.9m^2 (1,420sq ft)
Year entered service: 1943

Boeing B-29 Superfortress

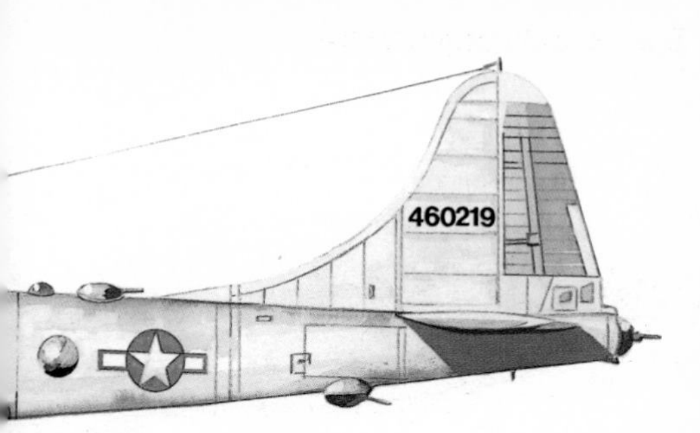

Country of Origin: United States
Type: strategic bomber and reconnaissance aircraft
Accommodation: normally ten
Armament: ten 13mm (.5in) and one 20mm (.75in) guns and 9,072kg (20,000lb) bomb load
Power Plant: four 2,200hp R-3350-79
Performance: maximum speed 586km/h (364mph) at 7,620m (25,000ft); service ceiling 9,750m (32,000ft); range 6,760km (4,200 miles)
Weights: empty 31,300kg (69,000lb); gross 62,370kg (137,500lb)
Dimensions: span 43.1m (141ft 3in); length 30.2m (99ft); height 9m (29ft 7in); wing area 161.3m²
(1,736sq ft)
Year entered service: 1943

Boulton Paul Defiant Mk II

Country of Origin: Great Britain
Type: fighter
Accommodation: two
Armament: four 8mm (.303in) Browning machine-guns in rear turret
Power Plant: one 1,030hp Rolls-Royce Merlin III
Performance: maximum speed 489km/h (304mph) at 5,180m (17,000ft); service ceiling 9,250m (30,350ft); range 748km (465 miles)
Weights: empty 2,760kg (6,078lb); loaded 3,900kg (8,600lb)
Dimensions: span 12m (39ft 4in); length 10.8m (35ft 4in); height 3.7m (12ft 2in)
Year entered service: 1940

Bristol Beaufighter Mk I

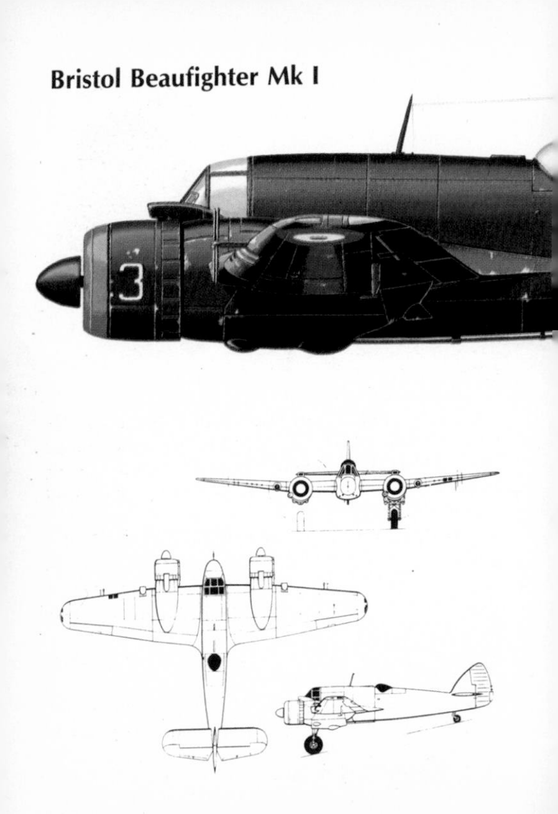

Country of Origin: Great Britain
Type: fighter
Accommodation: two
Armament: four 20mm cannon in belly; up to six 8mm (.303in) Browning machine-guns in wings; eight 7.5cm (3in) rockets on wing rail; one 728kg (1,605lb) torpedo, or up to 450kg (1,000lb) of bombs
Power Plant: two 1,600hp Bristol Hercules VI
Performance: maximum speed 530km/h (330mph) at sea level; service ceiling 8,840m (29,000ft); normal range 2,410km (1,500 miles); maximum range 2,820km (1,750 miles)
Weights: empty 6,380kg (14,070lb); maximum 9,435kg (21,100lb)
Dimensions: span 17.5m (57ft 10in); length 12.5m (41ft 4in); height 4.8m (15ft 10in); wing area 46.7m^2 (503sq ft)
Year entered service: 1940

Bristol Beaufort

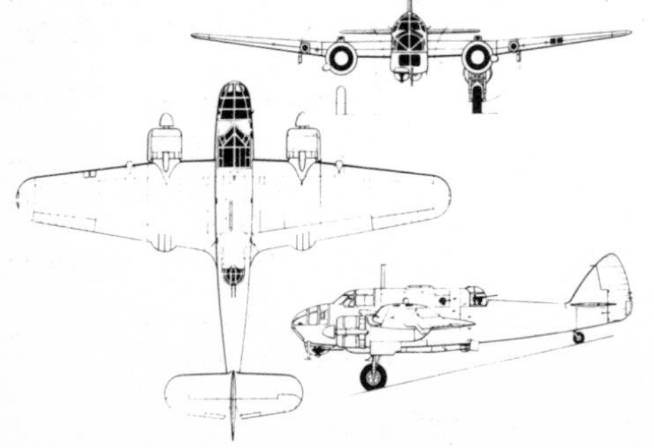

Country of Origin: Great Britain
Type: torpedo-bomber
Accommodation: four
Armament: two 8mm (.303in) guns each in nose and dorsal turret; maximum bomb load 680kg (1,500lb), or one 728kg (1,605lb) torpedo
Power Plant: two 1,130hp Bristol Taurus VI, XII or XVI
Performance: maximum speed 426km/h (265mph) at 1,830m (6,000ft); service ceiling 5,030m (16,500ft); range 2,570km (1,600 miles)
Weights: empty 5,940kg (13,100lb); maximum 9,630kg (21,230lb)
Dimensions: span 17.6m (57ft 10in); length 13.5m (44ft 3in); height 4.3m (14ft 3in); wing area 46.73m^2 (503sq ft)
Year entered service: 1940

Bristol Blenheim IV

Bristol Blenheim I

Country of Origin: Great Britain
Type: medium bomber
Accommodation: two
Armament: one 8mm (.303in) Browning machine-gun forward; one or two 8mm (.303in) VGO or Browning machine-guns in dorsal turret; four 8mm (.303in) Browning machine-guns in belly pack; maximum bomb load 635kg (1,400lb)
Power Plant: two 920hp Bristol Mercury XV
Performance: maximum speed 475km/h (295mph); service ceiling 9,600m (31,500ft); range 3,140km (1,950 miles)
Weights: empty 4,445kg (9,800lb); loaded 6,530kg (14,400lb)
Dimensions: span 17.2m (56ft 4in); length 13m (42ft 9in); height 3.9m (12ft 10in); wing area 43.6m^2 (469sq ft)
Year entered service: 1941

Chance-Vought F4U-1 Corsair

Country of Origin: United States
Type: naval fighter
Accommodation: one
Armament: four or six 13mm (.5in) machine-guns in wings
Power Plant: one 2,000hp Pratt & Whitney Double Wasp R-2800-8

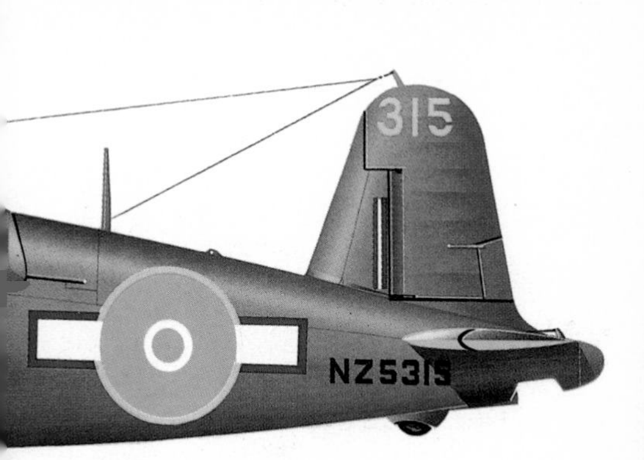

Performance: maximum speed 602km/h (374mph) at 7,010m (23,000ft); service ceiling 10,360m (34,000ft); maximum range 1,810km (1,125 miles)
Weights: empty 3,990kg (8,800lb); maximum loaded 6,350kg (14,000lb)
Dimensions: span 12.5m (41ft); length 10.2m (33ft 4in); height 4.6m (15ft 1in); wing area 29.2m^2 (314sq ft).
Year entered service: 1942

Commonwealth Boomerang

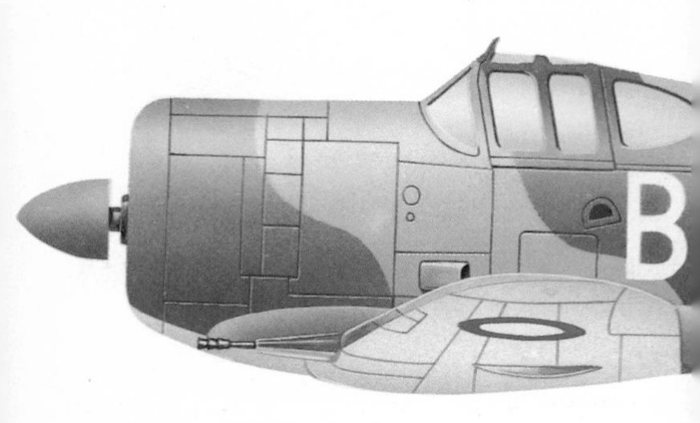

Country of Origin: Australia
Type: fighter
Accommodation: one
Armament: two 20mm Hispano cannon, and two 0.303in Browning machine-guns in each wing
Power Plant: one 1,200hp Pratt & Whitney R-1830 Twin Wasp 14-cylinder air-cooled radial

Performance: maximum speed 474km/h (296mph); service ceiling 8,845m (29,000ft); range 1,490km (930 miles)
Weights: empty 2,474kg (5,450lb); loaded 3,450kg (7,600lb)
Dimensions: span 11m (36ft 3in); length 7.77m (25ft 6in); height 3.5m (11ft 6in)
Year entered service: 1942

Designed as a stop-gap fighter for the Royal Australian Air Force after the Japanese attacked the American Pacific port of Pearl Harbor on 7 December 1941.

Consolidated B-24D Liberator

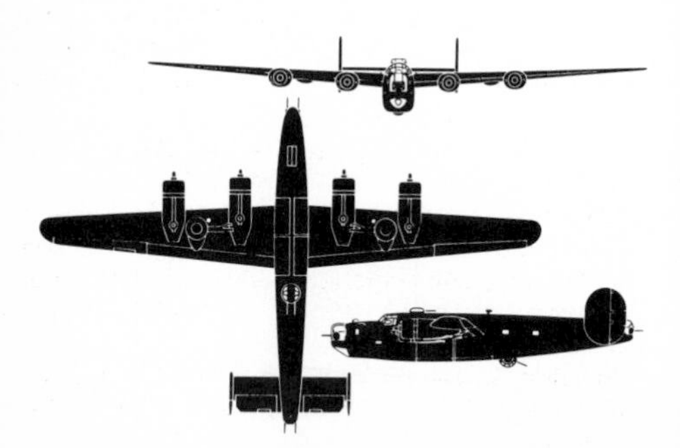

Consolidated B-24J Liberator

Country of Origin: United States
Type: heavy bomber
Accommodation: six to eight
Armament: twin 7.6mm (.3in), 8mm (.303in) or 13mm (.5in) machine-guns in nose, dorsal and tail turrets; single 8mm (.303m) or 13mm (.5in) machine-gun in waist positions; maximum bomb load 5,900kg (13,000lb)
Power Plant: four 1,200hp Pratt & Whitney Twin Wasp R-1830
Performance: maximum speed 430km/h (270mph) at 6,100m (20,000ft); service ceiling 9,750m (32,000ft); maximum range 3,700km (2,300 miles)
Weights: empty 16,780kg (37,000lb); loaded 28,120kg (62,000lb)
Dimensions: span 33.5m (110ft); length 20.4m (67ft 1in); height 5.5m (17ft 11in); wing area 97.4m^2 (1,048sq ft)
Year entered service: 1942

Consolidated PBY-5 Catalina

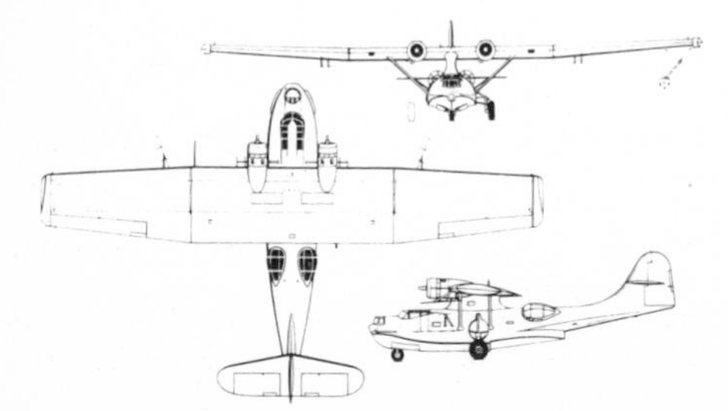

Country of Origin: United States
Type: patrol bomber flying-boat
Accommodation: seven–nine
Armament: one 13 mm (.5 in) machine-gun in each nose turret; two 7.6mm (.3in) Browning machine-guns in waist blisters; four depth-charges, two torpedoes or four 454kg (1,000lb) bombs
Power Plant: two 1,200hp Pratt & Whitney R-1830-82
Performance: maximum speed 322km/h (200mph); service ceiling 6,584m (21,600ft); range 3,050km (1,895 miles)
Weights: empty 7,893kg (17,400lb); maximum 15,145kg (33,390lb)
Dimensions: span 31.7m (104ft); length 19.5m (63ft 10in); height 5.76m (18ft 11in); wing area 130.1m^2 (1,400sq ft)
Year entered service: 1936

Curtiss P-40 Warhawk

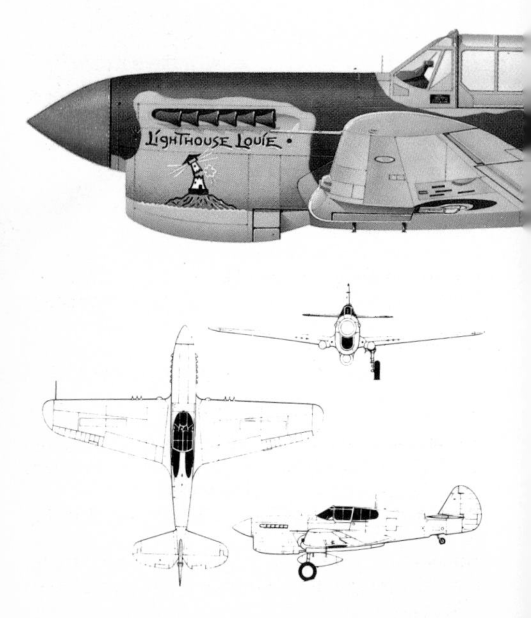

Curtiss P-40C Warhawk

Country of Origin: United States
Type: single-seat pursuit, ground-attack reconnaissance and advanced trainer
Accommodation: one
Armament: six 13mm (.5in) guns and one 227kg (500lb) bomb
Power Plant: one 1,360hp V-1710-81
Performance: maximum speed 608km/h (378mph) at 3,200m (10,500ft); service ceiling 11,582m (38,000ft); range 386km (240 miles)
Weights: empty 2,720kg (6,000lb); gross 4,014kg (8,850lb)
Dimensions: span 11.4m (37ft 4in); length 10.2m (33ft 4in); height 38m (12ft 4in); wing area $21.9m^2$ (236sq ft)
Year entered service: 1940

De Havilland Mosquito Mk XVI

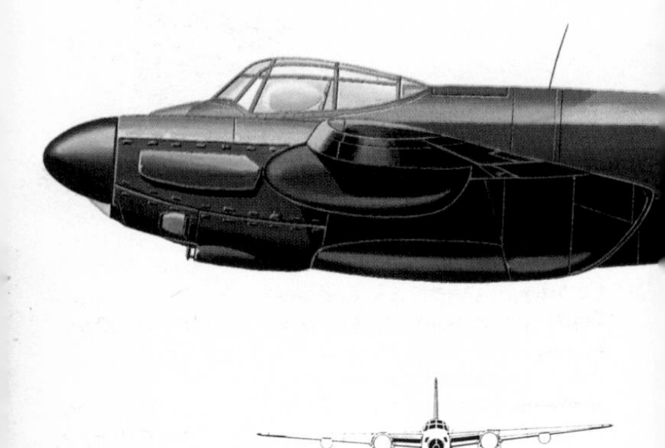

Country of Origin: Great Britain
Type: photo-reconnaissance (IX), fighter-bomber; bomber
Accommodation: two
Armament: four 20mm (.75in) Hispano cannon in belly (fighters); maximum bomb load 1,810kg (4,000lb) (bombers)
Power Plant: two 1,680hp Rolls-Royce Merlin 72
Performance: maximum speed 656km/h (408mph) at 7,925m (26,000ft); service ceiling 11,278m (37,000ft); maximum range 2,390km (1,485 miles)
Weights: empty 6,638kg (14,635lb); maximum 10,430kg (23,000lb)
Dimensions: span 16.5m (54ft 2in); length 13.6m (44ft 6in); height 3.8m (12ft 6in); wing area 42.2m^2 (454sq ft)
Year entered service: 1942

Dornier Do 17Z

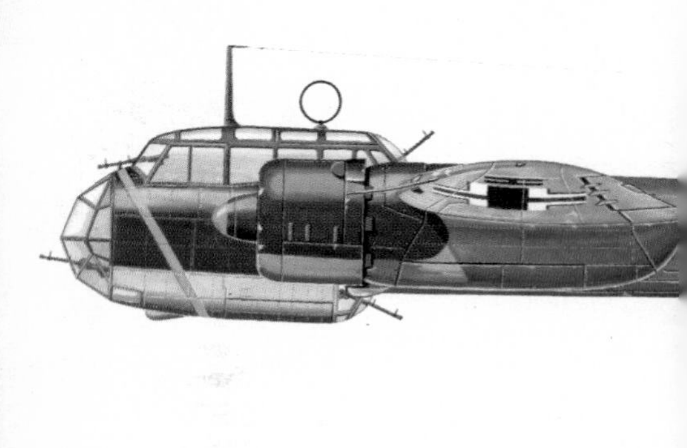

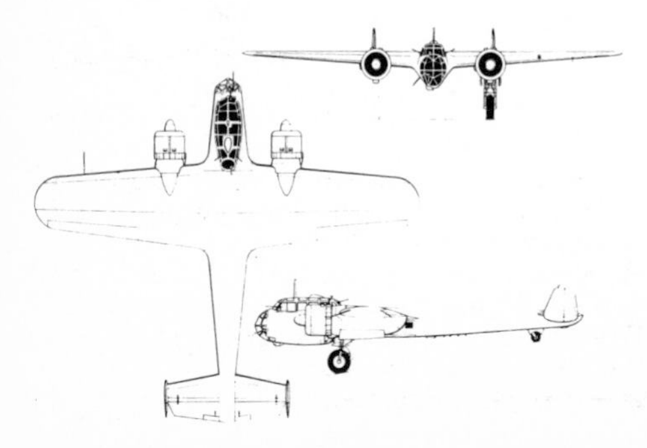

Country of Origin: Germany
Type: bomber
Accommodation: four
Armament: up to seven 7.9mm MG15 machine-guns; plus 1,000kg (2,205lb) bomb load
Power Plant: 2 1,000hp BMW-Bramo 323P nine-cylinder radial
Performance: maximum speed 425km/h (263mph); service ceiling 7,000m (22,960ft); range 1,160km (721 miles)
Weights: empty 5,209kg (11,484lb); maximum loaded 8,587kg (18,931)lb)
Dimensions: span 18m (59ft .5in); length 15.79m (51ft 9.5in); wing area 55m^2 (592 sq ft)
Year entered service: 1938

Dornier Do 217K

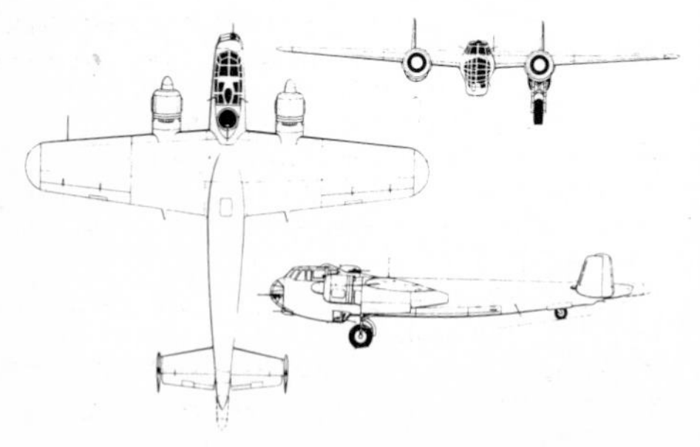

Dornier Do 217E-2

Country of Origin: Germany
Type: heavy night bomber
Accommodation: four
Armament: two 7.9mm MG81 machine-guns in nose, one 13mm MG131 machine-gun each in dorsal turret and ventral position; plus two (later four) 7.9mm MG81 machine-guns in rear cockpit lateral positions; provision for 4 L5 torpedoes; maximum bomb load 4,000kg (8,820lb)
Power Plant: two 1,700hp BMW 801 Ds
Performance: maximum speed 600km/h (348mph) at 5,700m (18,700ft); service ceiling 9,500m (31,180ft); maximum range 2,500km (1,555 miles)
Weights: empty 9,065kg (19,985lb); maximum 16,700kg (36,817lb)
Dimensions: span 19m (62ft 4in); length 17m (55ft 9.25in); height 4.96m (16ft 3.5in); wing area 57m^2 (613.5sq ft)
Year entered service: 1942

Douglas A-20G Havoc 'Boston'

Country of Origin: United States
Type: light bomber

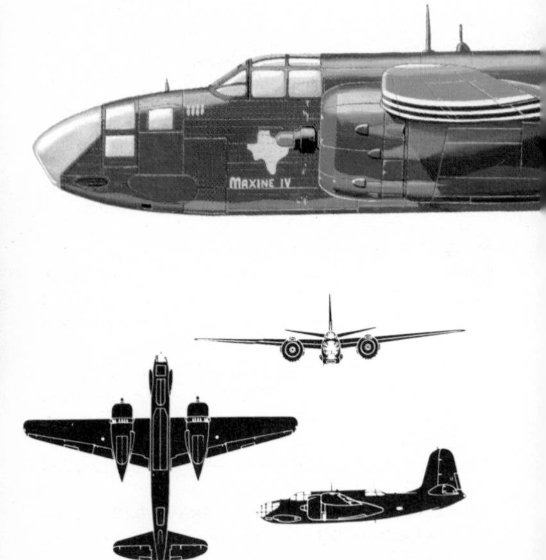

Douglas A-20C

Accommodation: two or three
Armament: eight 8mm (.303in) Browning machine-guns in nose (intruder: also one 8mm (.303in) Vickers K machine-gun in dorsal cockpit and up to 454kg/1,000lb bombs)
Power Plant: two 1,600hp Wright R-2600-23
Performance: maximum speed 545km/h (339mph) at 3,780m (12,400ft); service ceiling 7,860m (25,800ft); range 1,759km (1,090 miles).
Weights: empty 5,190kg (11,450lb); loaded 8,640kg (19,050lb)
Dimensions: span 18.7m (61ft 4in); length 14.3m (46ft 11.75in); height 4.8m (15ft 10in); wing area 43.1m^2 (464sq ft)
Year entered service: 1942

Douglas C-47 Skytrain 'Dakota'

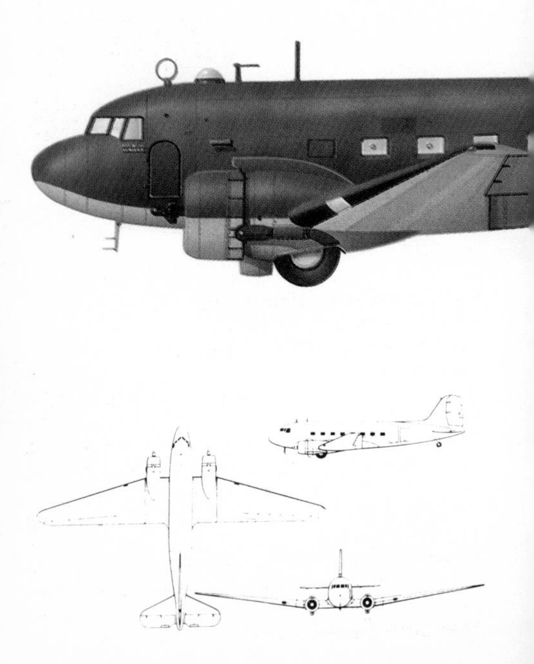

Country of Origin: United States
Type: transport
Accommodation: three, 28 passengers
Armament: nil
Power Plant: two 1,200hp Pratt & Whitney Twin Wasp R-1830-92
Performance: maximum speed 370km/h (230mph) at 2,590m (8,500ft); service ceiling 7,070m (23,200ft); maximum range 3,380km (2,100 miles)
Weights: empty 7,700kg (16,970lb); loaded 11,790kg (26,00lb)
Dimensions: span 29m (95ft); length 19.6m (64ft 5.5in); height 5.2m (16ft 11in); wing area 91.7m^2 (987sq ft)
Year entered service: 1941

Fairey Battle

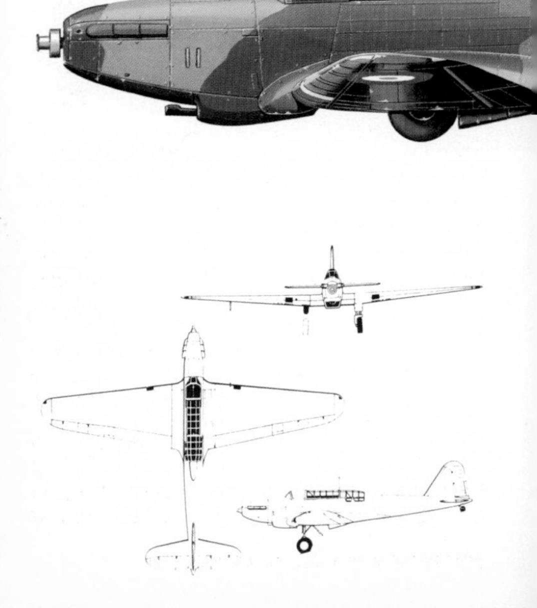

Country of Origin: Great Britain
Type: light bomber
Accommodation: three
Armament: one 8mm (.303in) Vickers K machine-gun in rear cockpit; one 8mm (.303in) Browning machine-gun forward; maximum bomb load 454kg (1,000lb)
Power Plant: one 1,030hp Rolls-Royce Merlin III
Performance: maximum speed 413km/h (257mph) at 6,100m (20,000ft); service ceiling 7,620m (25,000ft); range 1,610km (1,000 miles)
Weights: empty 3,015m (6,647lb); loaded 4,895kg (10,792lb)
Dimensions: span 16.5m (54ft); length 12.9m (42ft 4in); height 4.7m (15ft 6in); wing area 39.2m^2 (422sq ft)
Year entered service: 1937

Fairey Swordfish

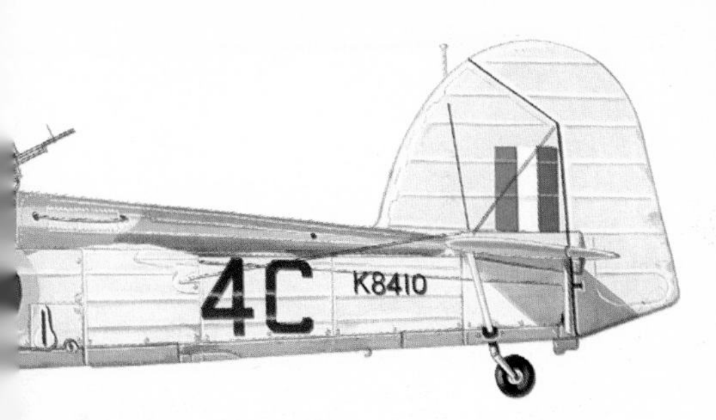

Country of Origin: Great Britain
Type: torpedo-bomber; maritime reconnaissance
Accommodation: three
Armament: one 8mm (.303in) Vickers K machine-gun in rear cockpit; one 730kg (1,610lb) torpedo, or up to 680kg (1,500lb) of bombs; six to eight 76mm (3in) rockets on underwing rails if required
Power Plant: one 690hp Bristol Pegasus III M3
Performance: maximum speed 248km/h (154 mph); service ceiling 5,867m (19,250ft); maximum range 1,610km (1,000 miles)
Weights: empty 1,903kg (4,195lb); loaded 3,500kg (7,720lb)
Dimensions: span 13.9m (45ft 6in); length 10.9m (35ft 8in); height 3.7m (12ft 4in); wing area 56.4m^2 (607sq ft)
Year entered service: 1936

Fiat G.50bis Freccia

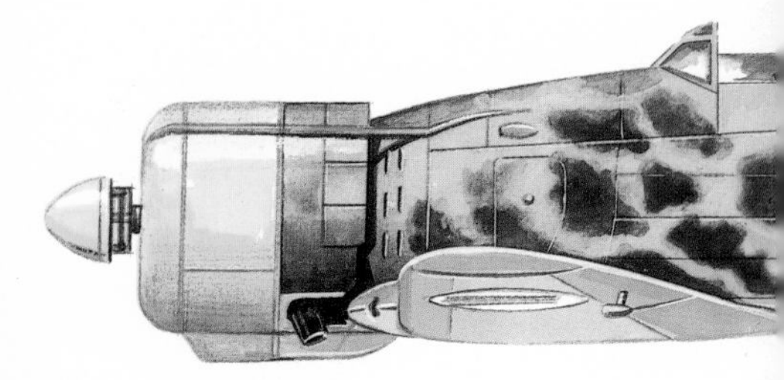

45 of these all metal monoplane fighters designed by Giuseppe Gabrielli were ordered by Regia Aeronautica in late 1937 for delivery the following year. A second production of some 200 were built with open cockpits due to problems with the sliding canopy, but the main production of over 400 of the G.50bis incorporated a reprofiled fuselage affording better pilot vision were built by CMASA.

Country of Origin: Italy
Type: fighter/fighter-bomber
Accommodation: one

Armament: two 13mm (.5in) Breda-S AFAT machine-guns in fuselage; maximum bomb load 300kg (660lb)
Power Plant: one 840hp Fiat A.74 RC38
Performance: maximum speed 470km/h (293mph) at 4,500m (14,765ft); service ceiling 10,700m (35,106ft); normal range 676km (420 miles)
Weights: empty 2,015kg (4,443lb); loaded 2,522kg (5,560lb)
Dimensions: span 10.9m (36ft 0.75in); length 7.8m (25ft 7in); height 2.95m (9ft 8.5in); wing area 18.25m^2 (196sq ft)
Year entered service: 1939

Focke-Wulf Fw 189A-1 Uhu

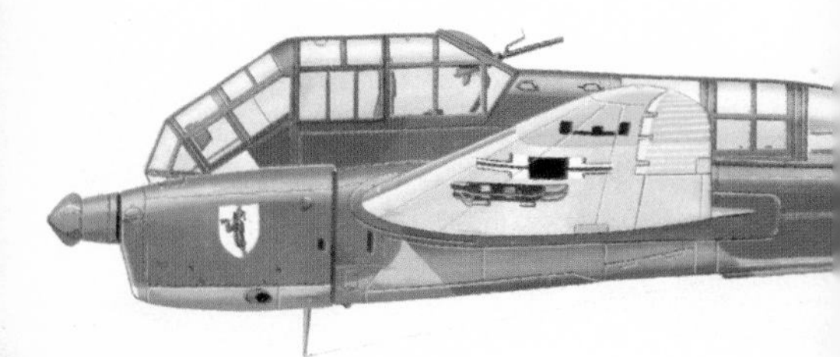

This extremely well constructed but unusual looking aircraft was known as 'The Flying Eye' when first seen in 1941. It had superb handling and performed beyond expectations on the Eastern front. Production over three versions topped 800 in French factories with final assembly at Bordeaux-Merignac which is today the Dassault-Breguet Mirage plant.

Country of Origin: Germany
Type: tactical reconnaissance liaison
Accommodation: three

Armament: two fixed forward-firing 7.9mm MG17 machine-guns in wing roots; plus one flexible 7.9mm MG15 machine-gun each in dorsal and fuselage tail cone positions; maximum bomb load 200kg (440lb)
Power Plant: two 465hp Argus As 410A-1
Performance: maximum speed 344km/h (214mph) at 2,500m (8,200ft); service ceiling 7,000m (22,970ft); maximum range 940km (584 miles)
Weights: empty 2,690kg (5,930lb); loaded 3,950kg (8,708lb)
Dimensions: wing span 18.4m (60ft 4.5in); length 11.9m (39ft 1in); height 3.1m (10ft 2in); wing area 38m^2 (409.3sq ft)
Year entered service: 1940

Focke-Wulf Fw 190

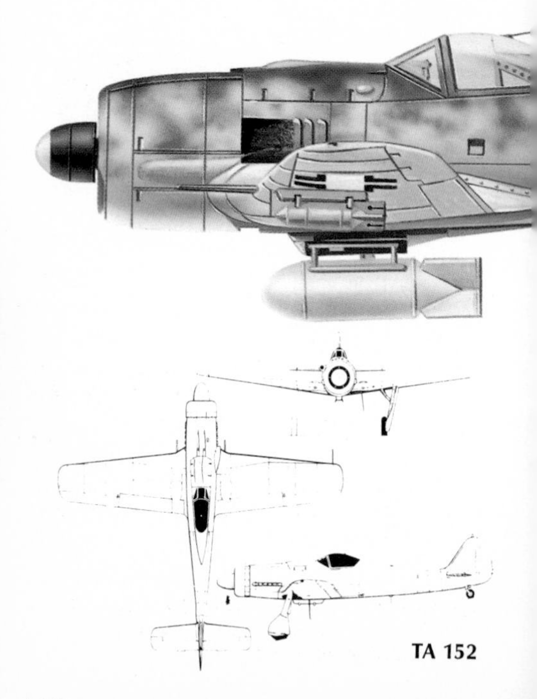

TA 152

Country of Origin: Germany
Type: fighter
Accommodation: one
Armament: two 7.9mm MG17 machine-guns above fuselage, two 20mm MG151 cannon in wing roots and two 20mm MGFF cannon outboard
Power Plant: one 1,700hp BMW 801D 18-cylinder two-row radial
Performance: maximum speed 650–675km/h (404–419mph); service ceiling 11,400m (37,400ft); range 800km (497 miles)
Weights: empty 3,471kg (7,652lb); maximum loaded 4,900kg (10,802lb)
Dimensions: span 10.5m (34ft 5.5in); length 8.8m (28ft 10.5in); wing area 18.3m^2 (196sq ft)
Year entered service: 1941

Focke-Wulf Fw 200C Condor

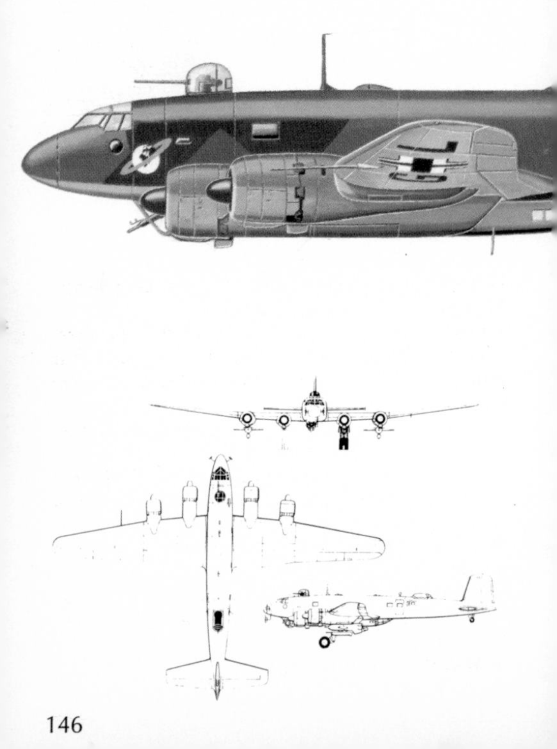

Country of Origin: Germany
Type: long-range maritime reconnaissance bomber
Accommodation: usually seven
Armament: three 7.9mm MG15 fired from turret behind flight deck, rear dorsal position and rear ventral hatch; four 250kg (551lb) bombs carried externally
Power Plant: four 830hp BMW 132H nine-cylinder radials
Performance: maximum speed 360km/h (224mph); service ceiling 5,790m (19,000ft); range 4,440km (2,759 miles)
Weights: empty 11,300kg (24,911lb); maximum loaded 17,500kg (38,580lb)
Dimensions: span 32.85m (107ft 9.5in); length 23.5m (76ft 11.75in); wing area 118.9m^2 (1,279sq ft)
Year entered service: 1938

Gloster Gladiator

Country of Origin: Great Britain
Type: fighter
Accommodation: one
Armament: two 8mm (.303in) Browning machine-guns in forward fuselage; two 8mm (.303in) Browning machine-guns under lower wings

Power Plant: one 840hp Bristol Mercury VIIIA
Performance: maximum speed 414km/h (257mph) at 4,450m (14,600ft); service ceiling 10,210m (33,500ft); normal range 690–710km (430–440 miles)
Weights: empty 1,459kg (3,217lb); loaded 2,083kg (4,592lb)
Dimensions: span 9.8m (32ft 3in); length 8.4m (27ft 5in); height 3.6m (11ft 9in); wing area 30m^2 (323sq ft)
Year entered service: 1937

Gloster Meteor I

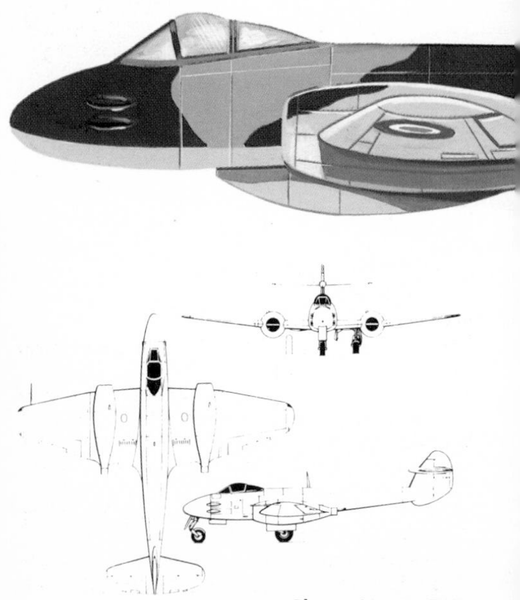

Gloster Meteor F.III

Country of Origin: Great Britain
Type: fighter
Accommodation: one
Armament: four fixed 20mm (.75in) British Hispano cannon in front fuselage
Power Plant: two 770kg (1,700lb) thrust Rolls-Royce W.2B/23C Welland Series 1 turbojets
Performance: maximum speed 668km/h (415 mph) at 3,050m (10,000ft); service ceiling 12,190m (40,000ft)
Weights: empty 3,690kg (8,140lb); loaded 6,260kg (13,795lb)
Dimensions: span 13.1m (43ft); length 12.6m (41ft 3in); height 4m (13ft); wing area 34.7m^2 (374sq ft)
Year entered service: 1944

Grumman F4F Wildcat

Country of Origin: United States
Type: naval fighter
Accommodation: one

Armament: six 13mm (.5in) machine-guns in wings
Power Plant: one 1,200hp Pratt & Whitney Twin Wasp R-1830-86
Performance: maximum speed 531km/h (330mph); service ceiling 8,534m (28,000ft); range 1,850km (1,150 miles)
Weights: empty 2,109kg (4,649lb); loaded 2,767kg (6,100lb)
Dimensions: span 11.6m (38ft); length 8.8m (28ft 9in); height 3.6m (11ft 10in); wing area 24.2m^2 (260sq ft)

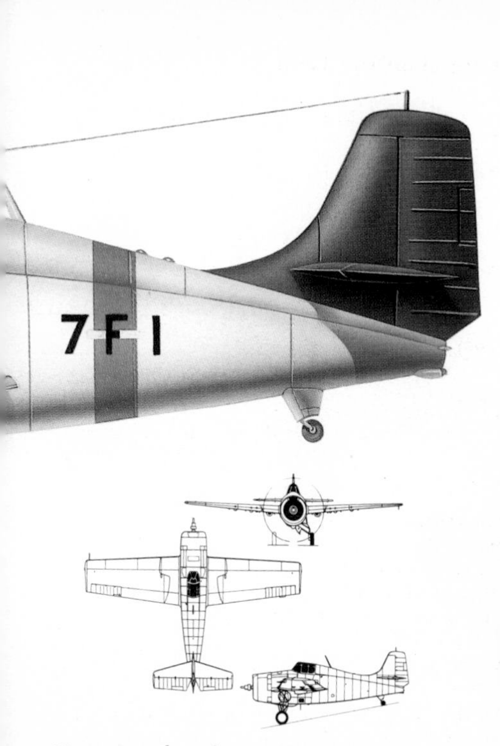

Year entered service: 1941

Grumman F6F Hellcat

Grumman F6F-3 Hellcat

Country of Origin: United States
Type: naval fighter
Accommodation: one
Armament: six 13mm (.5in) machine-guns in wings; up to 907kg (2,000lb) of bombs, or six 8cm (3in) rockets under wings
Power Plant: one 2,000hp Pratt & Whitney Double Wasp R-2800-10W
Performance: maximum speed 597km/h (371mph) at 5,240m (17,200ft); service ceiling 11,190m (36,700ft); maximum range 2,460km (1,530 miles)
Weights: empty 4,179kg (9,212lb); maximum loaded 6,238kg (13,753lb)
Dimensions: span 13.1m (42ft 10in); length 10.2m (33ft 7in); height 4m (13ft 1in); wing area $31m^2$ (334sq ft)
Year entered service: 1943

Grumman TBF Avenger

Country of Origin: United States
Type: torpedo bomber
Accommodation: two or three
Armament: two 13mm (.5in) machine-guns in wings; one 13mm (.5in) machine gun in dorsal turret; one 8mm (.303in) machine-gun in ventral location, one 871kg (1,921lb) 56cm (22in) torpedo, or up to 454kg (1,000lb) of bombs; eight 8cm (3in) rockets below wings if required

Power Plant: one 1,850hp Wright Cyclone GR-2600-8
Performance: maximum speed 417km/h (259mph) at 3,414m (11,200ft); service ceiling 7,010m (23,000ft); maximum range 3,074km (1,910 miles)
Weights: empty 4,810kg (10,600lb); loaded 7,390kg (16,300lb)
Dimensions: span 16.5m (54ft 2in); length 12.2m (40ft); height 4.8m (15ft 8in); 45.5m² (490sq ft)
Year entered service: 1942

Handley Page Halifax

Handley Page Halifax B.VI

Country of Origin: Great Britain
Type: heavy bomber
Accommodation: six–eight
Armament: four 8mm (.303in) Browning machine-guns in tail turret; two 8mm (.303in) Browning machine-guns in nose turret; one hand-operated 8mm (.303in) machine-gun in nose, two in beam locations; four 8mm (.303in) Browning machine-guns in dorsal turret; maximum bomb load 5,900kg (13,000lb)
Power Plant: four 1,615hp Bristol Hercules XVI
Performance: maximum speed 454km/h (282mph) at 4,120m (13,500ft); service ceiling 7,315m (24,000ft); normal range 3,194km (1,985 miles)
Weights: empty 17,346kg (38,240lb); maximum loaded 29,480kg (65,000lb)
Dimensions: span 30.1m (98ft 10in); length 21.8m (71ft 7in); height 6.3m (20ft 9in); wing area116.1m^2 (1,250sq ft)
Year entered service: 1940

Handley Page Hampden

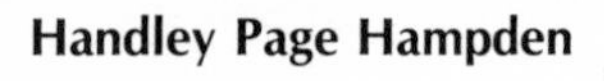

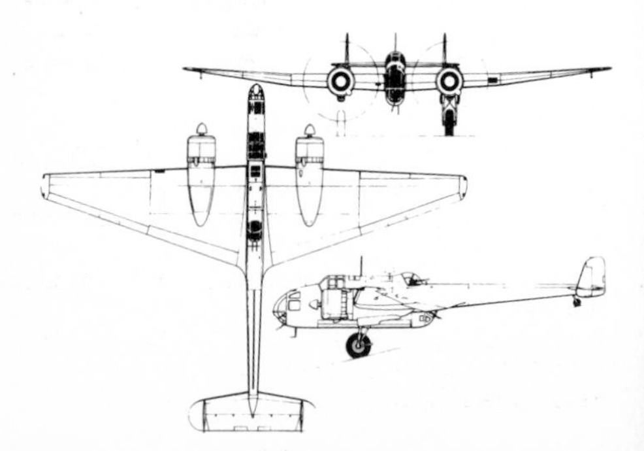

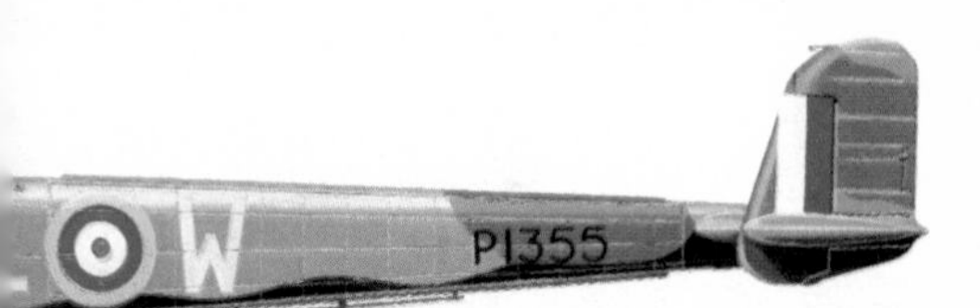

Country of Origin: Great Britain
Type: medium day bomber
Accommodation: four
Armament: one 8mm (.303in) Vickers K machine-gun in top forward fuselage; one 8mm (.303in) Vickers K machine-gun in nose, one or two 8mm (.303in) Vickers K machine-guns in dorsal cupola; one or two 8mm (.303in) Vickers K machine-guns in ventral cupola; maximum bomb load 1,810kg (4,000lb)
Power Plant: two 980hp Bristol Pegasus XVIII
Performance: maximum speed 426km/h (265mph) at 4,724m (15,500ft); service ceiling 6,920m (22,700ft); range with 1,810kg (4,000lb) bomb load 1,400km (870 miles)
Weights: empty 5,343kg (11,780lb); normal loaded 8,507kg (18,756lb)
Dimensions: span 21.1m (69ft 2in); length 16.3m (53ft 7in); height 4.5m (14ft 11in)
Year entered service: 1938

Hawker Hurricane I

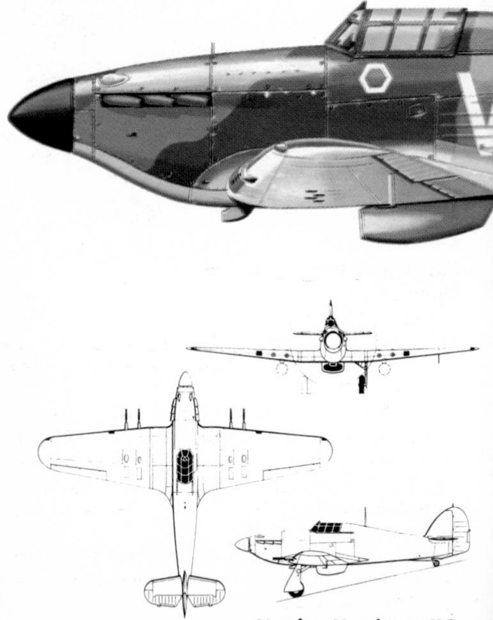

Hawker Hurricane IIC

Country of Origin: Great Britain
Type: fighter
Accommodation: one
Armament: eight 8mm (.303in) Browning machine-guns in wings; up to 454kg (1,000lb) of bombs underwing
Power Plant: one 1,030hp Rolls Royce Merlin II
Performance: maximum speed 496km/h (308mph) at 3,048m (10,000ft); service ceiling 10,180m (33,400ft); maximum range 845km (525 miles)
Weights: empty 2,151kg (4,743lb); loaded 2,820kg (6,218lb)
Dimensions: span 12.2m (40ft); length 9.5m (31ft 4in); height 4m (13ft 4.5in); wing area 24m^2 (259sqft)
Year entered service: 1937

Hawker Tempest

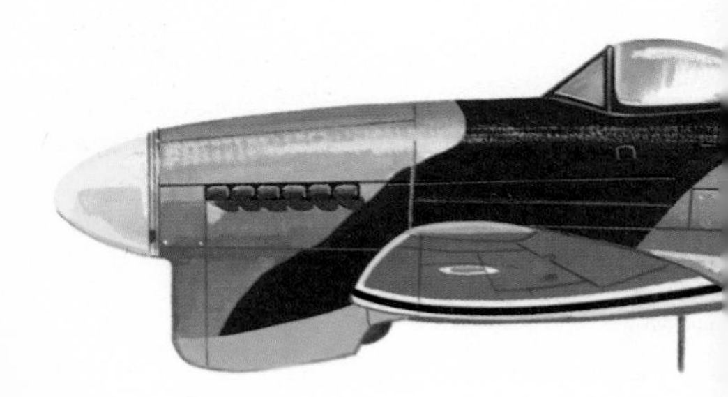

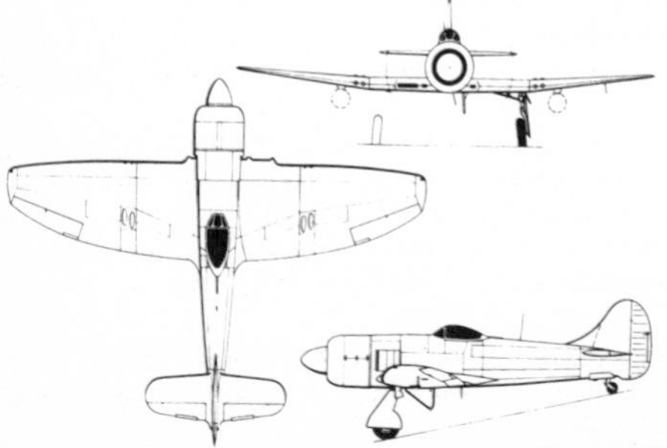

Tempest II

Country of Origin: Great Britain
Type: fighter-bomber
Accommodation: one
Armament: four 20mm Hispano cannon in wings; up to 907kg (2,000lb) of bombs
Power Plant: one 2,180hp Napier Sabre IIA, B or C
Performance: maximum speed 686km/h (426mph) at 5,639m (18,500ft); service ceiling 11,125m (36,500ft); normal range 1,191km (740 miles)
Weights: empty 4,080kg (9,000lb); loaded 6,100kg (13,450lb)
Dimensions: span 12.5m (41ft); length 10.26m (34ft 5in); height 4.9m (16ft 1in); wing area 28.1m² (302sqft)
Year entered service: 1944

Hawker Typhoon

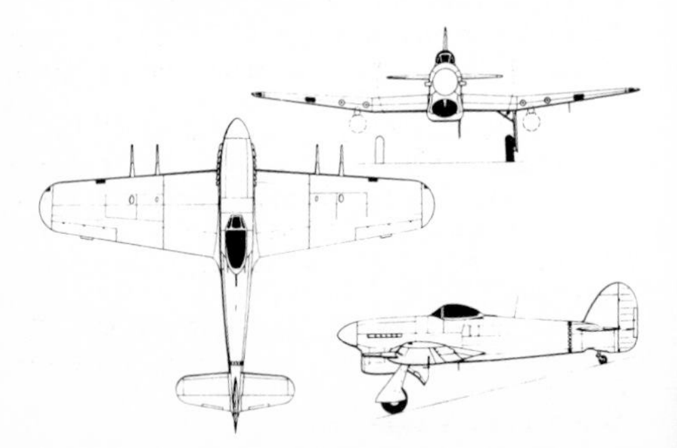

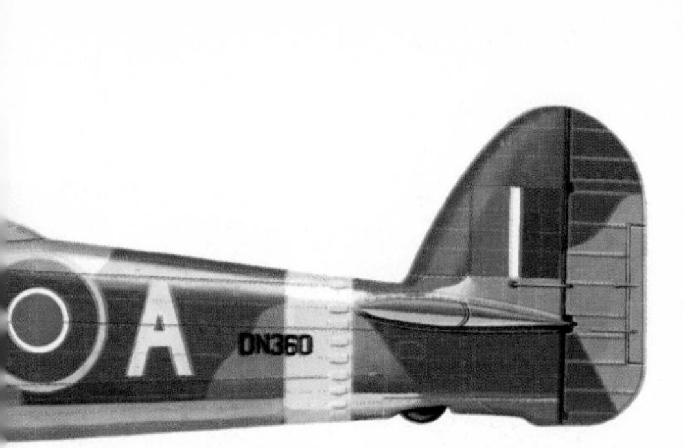

Country of Origin: Great Britain
Type: fighter
Accommodation: one
Armament: four 20mm Hispano cannon; provision for eight 8cm (3in) rocket projectiles or two 454kg (1,000lb) bombs underwing
Power Plant: one 2,260hp Napier Sabre IIC
Performance: maximum speed 663km/h (412mph) at 5,790m (19,000ft); service ceiling 10,730m (35,200ft); maximum range 820km (510 miles)
Weights: empty 4,010kg (8,840lb); maximum 6,340kg (13,980lb)
Dimensions: span 12.7m (41ft 7in); length 9.7m (31ft 11.5in); height 4.7m (15ft 4in); wing area 25.9m² (279sq ft)
Year entered service: 1943

Heinkel He 111

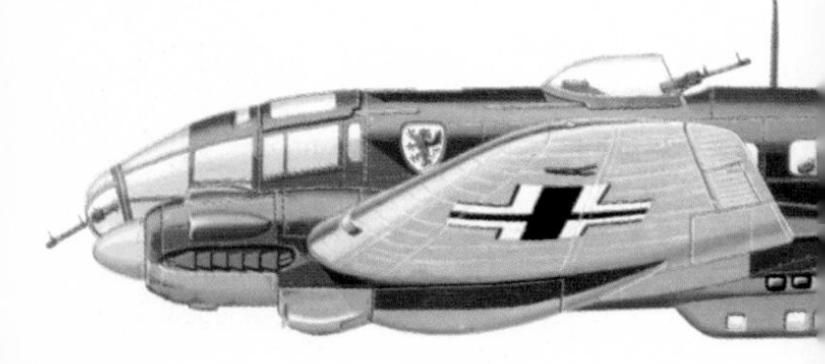

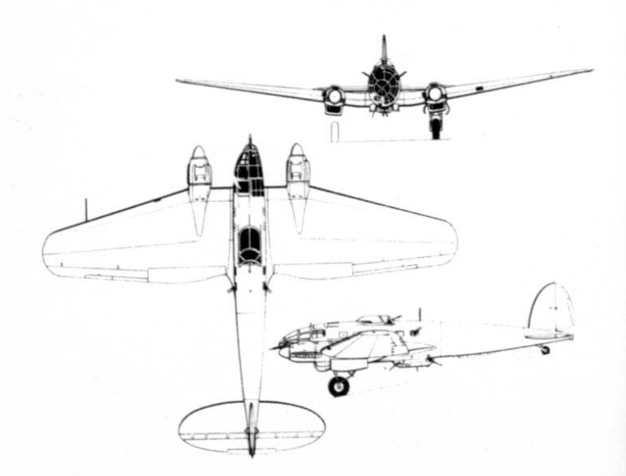

Country of Origin: Germany
Type: medium bomber
Accommodation: four
Armament: three 7.9mm MG15 machine-guns aimed from nose, dorsal position and retractable ventral 'dustbin'; bomb load 2,000kg (4,410lb)
Power Plant: two 1,010hp Junkers Jumo 211A-1 inverted vee-12
Performance: maximum speed 415km/h (258mph); service ceiling 7,000m (22,970ft); range 1,950km (1,212 miles)
Weights: empty 6,818kg (15,031lb); maximum loaded 9,600kg (21,164lb)
Dimensions: span 22.6m (74ft 1.75in); length 17.5m (57ft 5in); wing area 87.6m^2 (942.9sq ft)
Year entered service: 1936

Heinkel He 219A-5/R2 Uhu

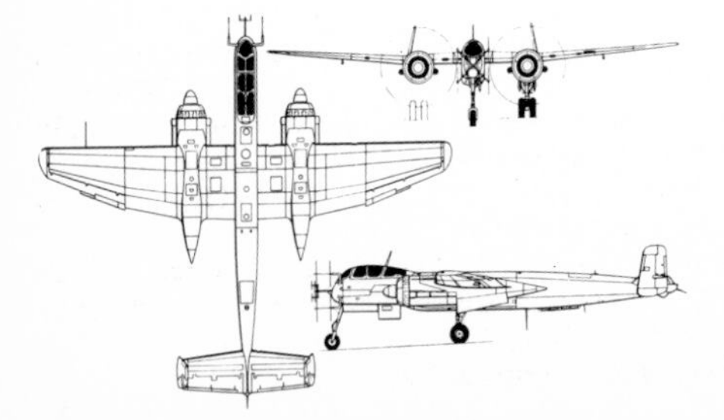

Country of Origin: Germany
Type: night-fighter
Accommodation: two
Armament: two 20mm MC151 cannon in wings; two 30mm MK108 cannon in ventral tray; two oblique upward-firing 30mm MK108 cannon in rear fuselage
Power Plant: two 1,800hp Daimler-Benz DB603E
Performance: maximum speed 630km/h (391mph) at sea level; service ceiling 11,300m (37,070ft); maximum range 2,800km (1,740 miles)
Weights: empty 9,900kg (21,826lb); loaded 13,150kg (28,990lb)
Dimensions: span 18.5m (60ft 8.3in); length 15.5m (50ft 11.75in); height 4.1m (13ft 5.5in); wing area 44.5m^2 (479sq ft)
Year entered service: 1943

Ilyushin Il-2M3

Country of Origin: Soviet Union
Type: ground-attack
Accommodation: two

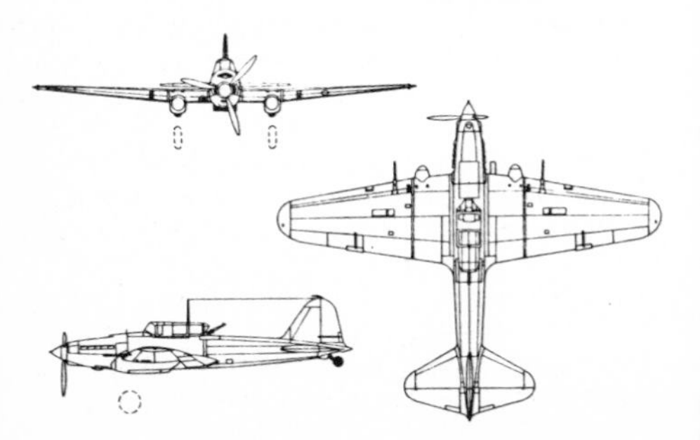

Armament: two fixed forward-firing 7.62mm ShKAS machine-guns in forward fuselage; maximum bomb load 600kg (1,321lb) or eight 82mm RS 82 (or 132mm RS 132 rocket projectiles)
Power Plant: one 1,750hp Mikulin AM-38F
Performance: maximum speed 425km/h (264mph); service ceiling 6,500m (21,325ft); range 765km (475 miles)
Weights: empty 3,400kg (7,495lb); maximum loaded 5,872kg (12,945lb)
Dimensions: span 14.6m (47ft 10.75in); length 11.6m (38ft .5in); height 3.4m (11ft 2in), wing area 38.5m² (414.4sq ft)
Year entered service: 1941

Ilyushin Il-10

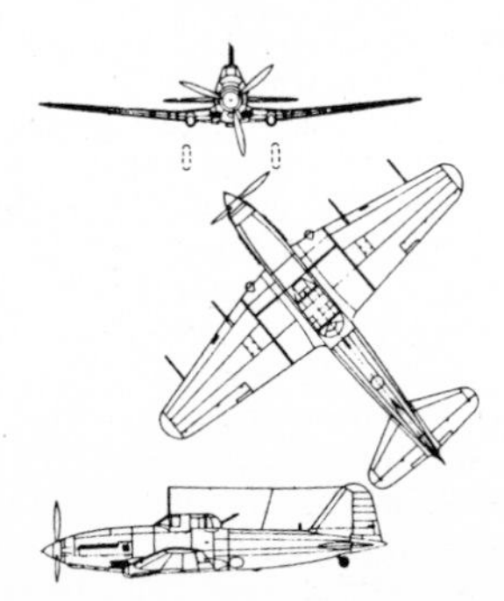

Country of Origin: Soviet Union
Type: ground-attack
Accommodation: two
Armament: two 7.62mm ShKAS machine-guns and two 23mm NS-23 cannon in wings, plus one 20mm cannon in dorsal cupola; provision for 400kg (880lb) bomb load, plus eight 82mm RS 82 rocket projectiles
Power Plant: one 2,000hp Mikulin AM-42
Performance: maximum speed 507km/h (315mph) at sea level; service ceiling 7,500m (24,606ft); range 1,000km (620 miles).
Weights: empty 4,500kg (9,920lb); maximum loaded 6,536kg (14,409lb)
Dimensions: span 13.4m (43ft 11.5in); length 11.2m (36ft 9in); wing area 30m^2 (323sq ft)
Year entered service: 1945

Junkers Ju 52/3M G7E

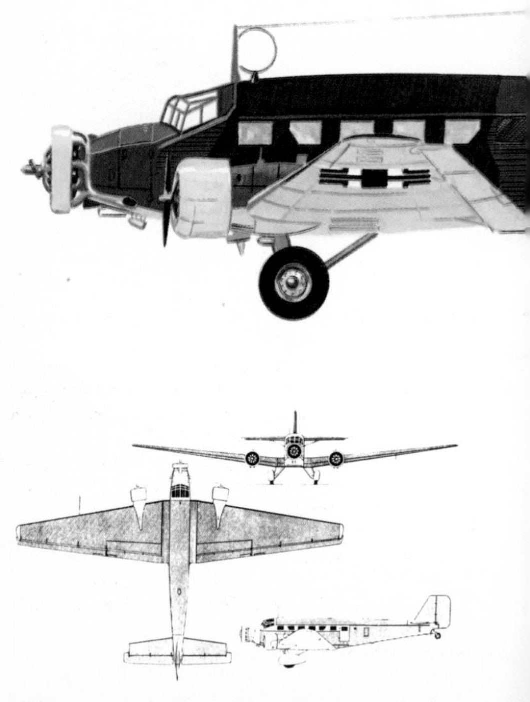

Country of Origin: Germany
Type: medium transport
Accommodation: three plus 18 troops or 12 stretchers
Armament: one 7.9mm MG15 machine-gun in open aft dorsal position; one 7.9mm MG15 machine-gun each in forward upper and two beam positions
Power Plant: three 830hp BMW 132T-2
Performance: maximum speed 272km/h (169mph) at sea level; service ceiling 5,500m (18,000ft); range 1,500km (930 miles).
Weights: empty 6,500kg (14,328lb); loaded 11,030kg (24,320lb)
Dimensions: span 29.2m (95ft 10in); length 18.9m (62ft); height 4.52m (14ft 10in); wing area 110.5m^2 (1,190sq ft)
Year entered service: 1936

Junkers Ju 87B-1 Stuka

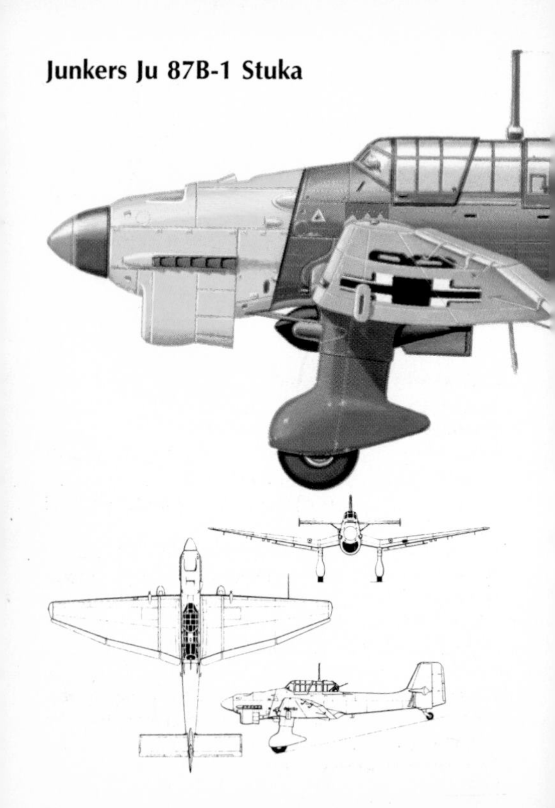

Country of Origin: Germany
Type: dive-bomber
Accommodation: two
Armament: two fixed forward-firing 7.9mm MG17 machine-guns in wings; one flexible 7.9mm MG15 machine-gun in rear cockpit; maximum bomb load 500kg (1,102lb)
Power Plant: one 1,200hp Junkers Jumo 211Da
Performance: maximum speed 340km/h (211mph); service ceiling 8,000m (26,150ft); maximum range 790km (490 miles)
Weights: empty 2,710kg (5,980lb); maximum 4,340kg (9,560lb)
Dimensions: span 13.8m (45ft 3.3in); length 11.1m (36ft 5in); height 4m (13ft 2in); wing area 31.9m^2 (343sq ft)
Year entered service: 1938

Junkers Ju 88D–1

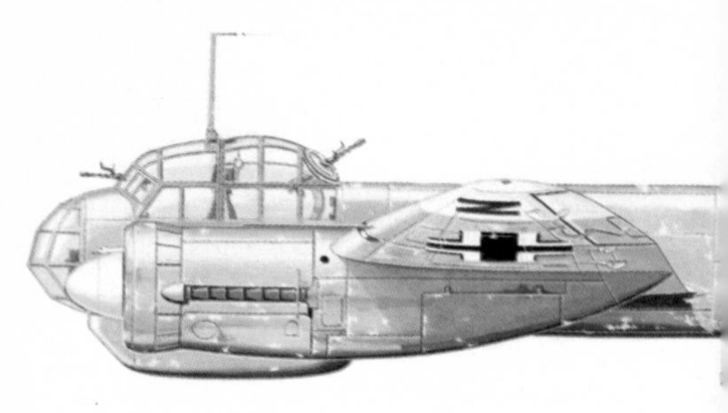

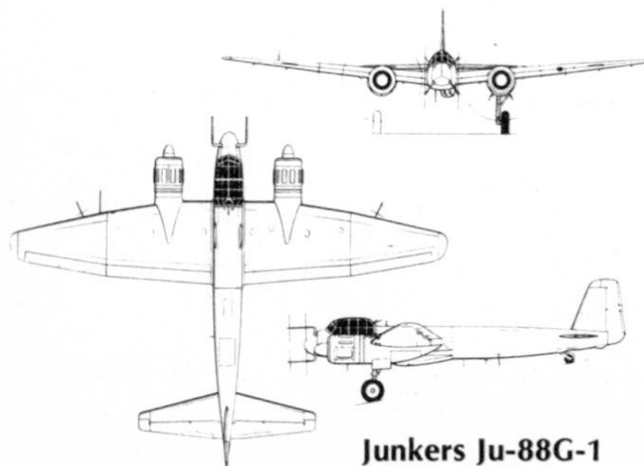

Junkers Ju-88G-1
Night fighter version

Country of Origin: Germany
Type: long-range reconnaissance
Accommodation: four
Armament: one 7.9mm MG15 machine-gun each in nose, dorsal, and ventral positions
Power Plant: two 1,340hp Junkers Jumo 211J
Performance: maximum speed 485km/h (301mph) at 4,800m (15,750ft); service ceiling 8,000m (26,250ft); normal range 4,800km (2,980 miles)
Weights: empty 8,850kg (19,510lb); loaded 12,350kg (27,230lb)
Dimensions: span 20.1m (65ft 10.5in); length 14.4m (47ft 1.3in); height 4.8m (15ft 9in); wing area 54.5m^2 (586.6sq ft)
Year entered service: 1940

Lavochin La-5FN

Country of Origin: Soviet Union
Type: fighter

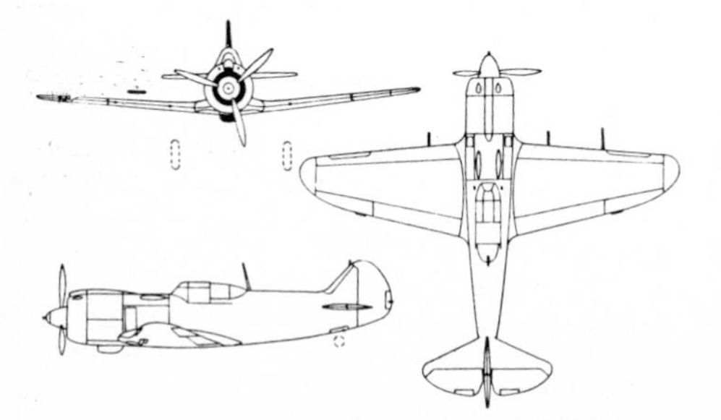

Accomodation: one
Armament: two 20mm SHAVAK cannons in upper engine cowling
Power Plant: one 1,850hp fuel-injected Shvetsov M-82FN radial
Performance: maximum speed 550km/h (342mph) at sea level; service ceiling 10,000m (32,000ft); range 785km (475 miles)
Weights: empty 2,800kg (6,173lb);
loaded 3,360kg (7,407lb)
Dimensions: span 9.8m (32ft 2in); length 8.6m (28ft 2in); height 2.54m (8ft 4in); wing area 17.5m^2 (188.4sq ft)
Year Entered Service: 1942

Lavochkin La-7

Country of Origin: Soviet Union
Type: interceptor fighter

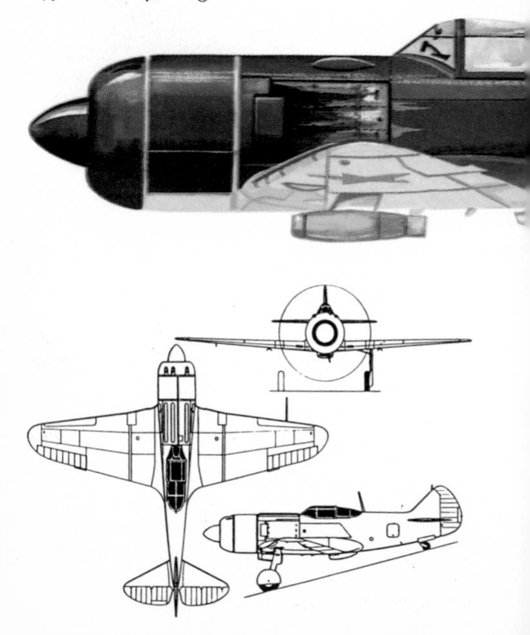

Accommodation: one
Armament: two or three 20mm Beresin B-20 cannon in upper cowling; provision for 200kg (440lb) bombs underwing
Power Plant: one 1,850hp Shvetsov M-82FN
Performance: maximum speed 665km/h (413mph); service ceiling 10,000m (34,450ft); range 635km (395 miles)
Weights: empty 2,638kg (5,816lb); normal 3,400kg (7,496lb)
Dimensions: span 9.8m (32ft 1.75in); length 8.6m (28ft 2.5in); wing area 17.5m² (188.4sq ft)
Year entered service: 1943

Lockheed P-38 Lightning

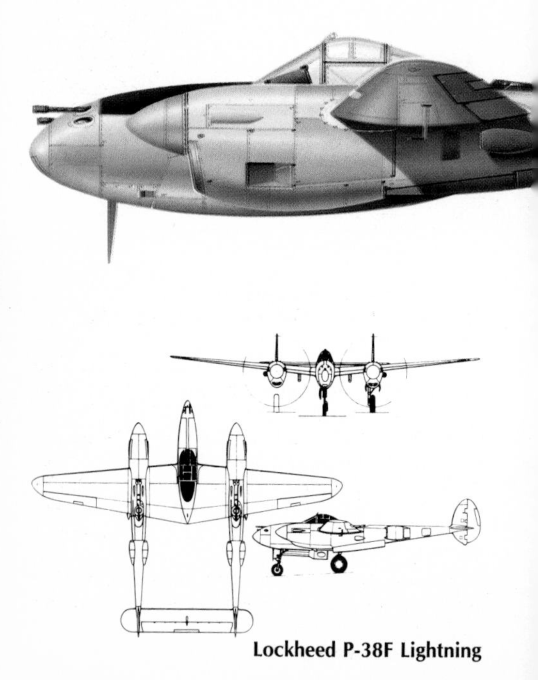

Lockheed P-38F Lightning

Country of Origin: United States
Type: single seat pursuit and long-range fighter
Accommodation: one
Armament: one 20mm Hispano cannon and four 13mm (.5in) Browning machine-guns in nose; maximum bomb load 907kg (2,000lb)
Power Plant: two 1,250hp Allison V-1710-49/53
Performance: maximum speed 558km/h (347mph) at 1,520m (5,000ft); service ceiling 11,890m (39,000ft); maximum range 2,293km (1,425 miles)
Weights: empty 6,169kg (13,600lb); maximum 9,070kg (20,000lb)
Dimensions: span 15.85m (52ft); length 11.5m (37ft 10in); height 3m (9ft 10in); wing area 30.4m^2 (327.5sq ft)
Year entered service: 1941

Macchi MC. 200 Saetta

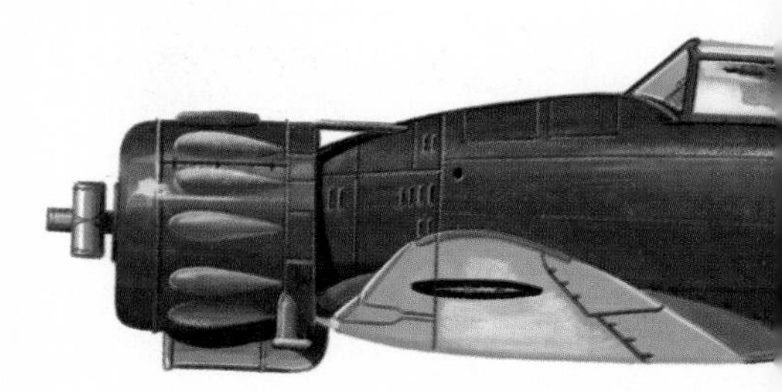

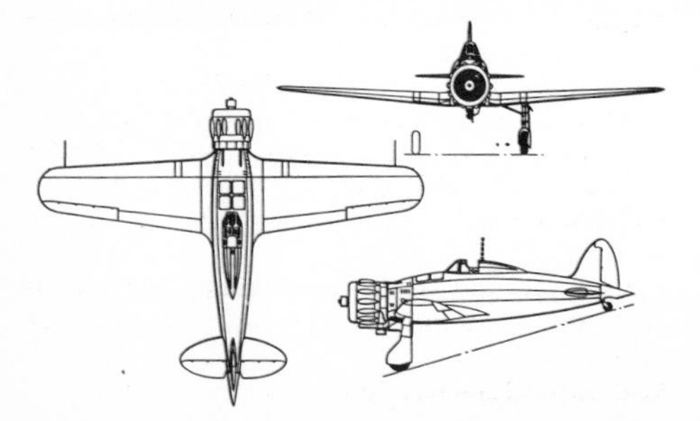

Country of Origin: Italy
Type: interceptor fighter
Accommodation: one

Armament: two 12.7mm Breda-SAFAT machine-guns in upper cowling
Power Plant: one 870hp Fiat A.74 RC 38
Performance: maximum speed 502km/h (312mph) at 4,500m (14,765ft); service ceiling 8,900m (29,200ft); maximum range 870km (540 miles)
Weights: empty 1,800kg (3,900lb); loaded 2,200kg (4,850lb)
Dimensions: span 10.6m (34ft 8.6in); length 8.2m (26ft 10.5in); height 3.5m (11ft 6.25in); wing area 16.8m^2 (181sq ft)
Year entered service: 1939

Macchi MC.205V Veltro

Country of Origin: Italy
Type: fighter/fighter bomber
Accommodation: one

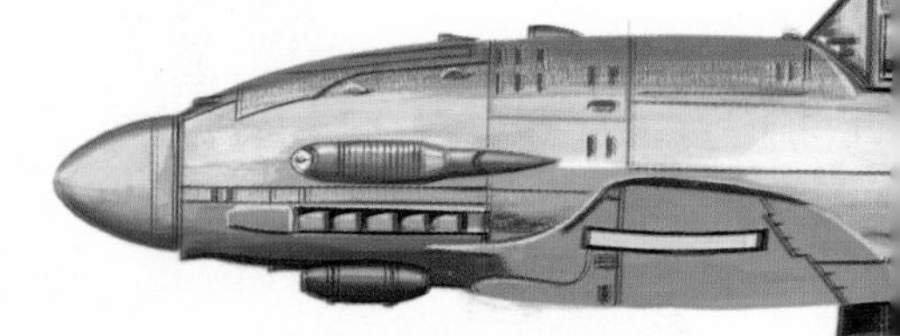

Armament: two 12.7mm Breda-SAFAT machine-guns in upper cowling; two 7.7mm Breda-SAFAT machine-guns in wings; maximum bomb load 320kg (700lb)
Power Plant: one 1,475hp Fiat RA.1050 RC 58 Tifone
Performance: maximum speed 642km/h (399mph) at 7,200m (23,620ft); service ceiling 11,000m (36,090ft); normal range 1,040km (646 miles)
Weights: empty 2,524kg (5,564lb); loaded 3,224kg (7,108lb)
Dimensions: span 10.6m (34ft 8.5in); length 8.85m (29ft .5in); height 3.02m (9ft 10.75in); wing area 16.8m^2 (180.8sq ft)
Year entered service: 1943

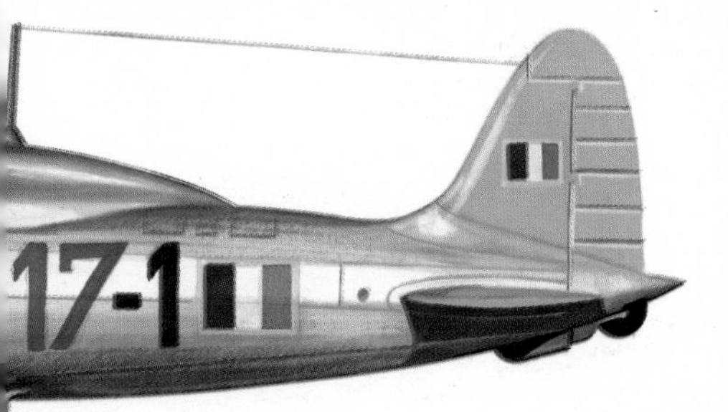

First flown on 19 April 1942 though production did not become available until June 1943 when only a limited number were produced. A modified version of the MC202 it came too late to have much impact, but in that short time it gained a reputation as the best Italian fighter aircraft of World War II and matched the best of the USA fighters such as the North American P-51D Mustang. The fighter was also used by the Luftwaffe.

Martin B-10

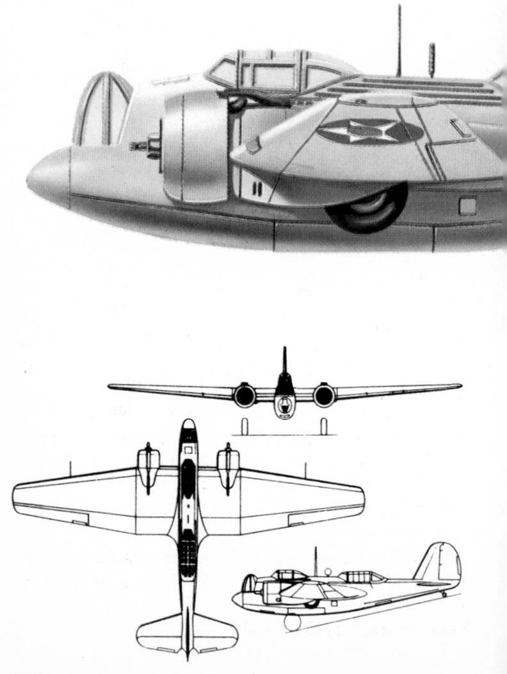

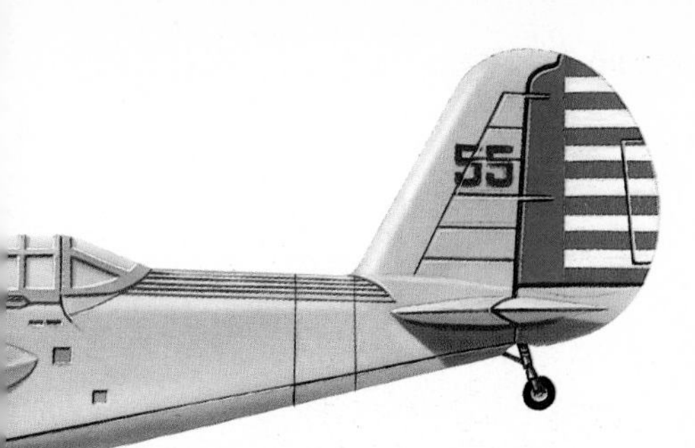

Country of Origin: United States
Type: light bomber
Accommodation: four
Armament: one 8mm (.303in) Browning machine-gun in nose and rear turrets and one in ventral turret; up to 1,025kg (2,260lb) bombs
Power Plant: two 775hp Wright R-1820-33
Performance: maximum speed 343km/h (213mph); service ceiling 7,376m (24,200ft); range 1,995km (1,240 miles)
Weights: empty 4,390kg (9,681lb); gross 7,440kg (16,400lb)
Dimensions: span 21.5m (70ft 6in); length 13.6m (44ft 9in); height 4.7m (15ft 5in); wing area 63m^2 (678sq ft)
Year entered service: 1934

Martin B-26B Marauder

Country of Origin: United States
Type: medium bomber
Accommodation: six or seven

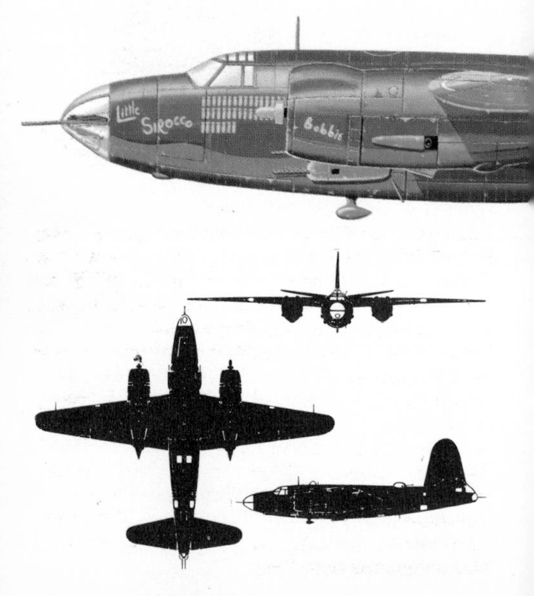

Armament: two 13mm (.5in) machine-guns in nose; two 13mm (.5in) machine guns in dorsal turret; two 13mm (.5in) machine guns in tail turret, single 13mm (.5in) machine-gun in each beam location; maximum bomb load 2,360kg (5,200lb)
Power Plant: two 2,000hp Pratt & Whitney Double Wasp R-2800-43
Performance: maximum speed 510km/h (317mph) at 4,420m (14,500ft); service ceiling 8,530m (28,000ft); range 1,850km (1,150 miles)
Weights: empty 7,710kg (17,000lb); loaded 16,780kg (37,000lb)
Dimensions: span 21.6m (71ft); length 17.5m (57ft 6in); height 6.1m (20ft); wing area 55.9m^2 (602sq ft)
Year entered service: 1942

Messerschmitt Bf 109G-2

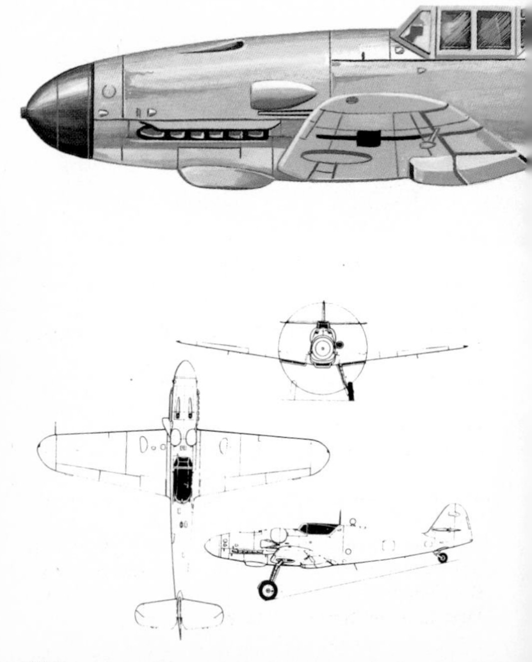

Country of Origin: Germany
Type: fighter
Accomodation: one
Armament: one 20mm MG151/20 cannon firing through propeller hub
Power Plant: one 1,475hp Daimler-Benz DB 605A
Performance: maximum speed 569km/h (353mph) at sea level; service ceiling 12,000m (39,370ft); maximum range 850km (528 miles)
Weights: empty 2,253kg (4,968lb); maximum 3,200kg (7,055lb)
Dimensions: span 9.9m (32ft 6.5in); length 8.85m (29ft 5in); height 2.5m (8ft 2.5in); wing area 16.1m^2 (173.3sq ft)
Year Entered Service: 1942

Messerschmitt Bf 110F-2

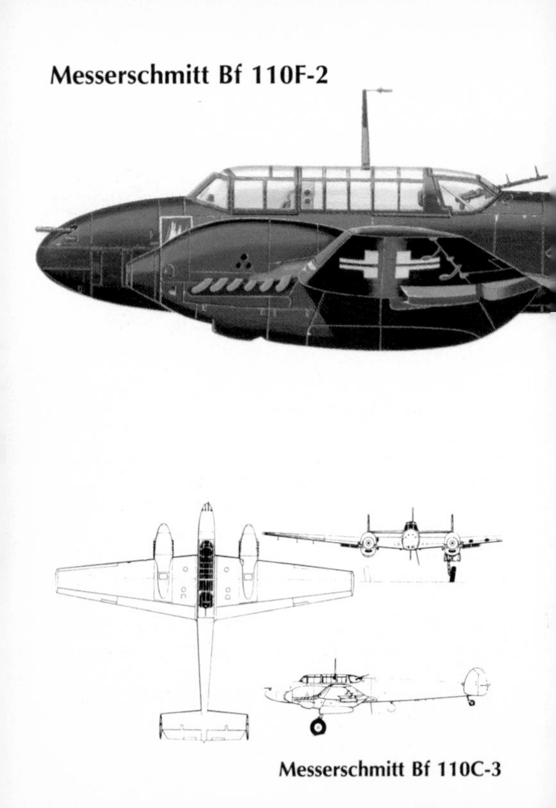

Messerschmitt Bf 110C-3

Country of Origin: Germany
Type: fighter
Accommodation: two–three
Armament: four fixed forward-firing 20mm MG FF cannon in nose; plus one flexible 7.9mm MG15 machine-gun in rear cockpit
Power Plant: two 1,100hp Daimler Benz DB 601A-1
Performance: maximum speed 525km/h (326mph) at 4,000m (13,120ft); service ceiling 10,000m (32,810ft); maximum range 1,100km (680 miles)
Weights: empty 4,425kg (9,755lb); maximum 6,750kg (14,880lb)
Dimensions: span 16.25m (53ft 3.75in); length 12.7m (39ft 7.25in); height 4.13m (13ft 6.5in); wing area 38.4m^2 (413.3 sq ft)
Year entered service: 1940

Messerschmitt Me 262A-1a

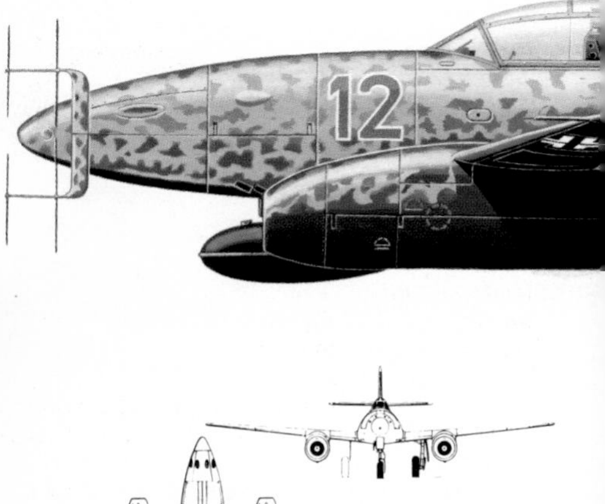

Country of Origin: Germany
Type: interceptor fighter
Accommodation: one
Armament: four 30mm Mk 108 cannon in nose
Power Plant: two 900kg (1,984lb) thrust Junkers Jumo 004B-1, B-2 or B-3
Performance: maximum speed 870km/h (540mph) at 6,000m (19,690ft); service ceiling 11,450m (37,570ft); normal range 845km (525 miles)
Weights: empty 3,800kg (8,378lb); loaded 6,400kg (14,108lb)
Dimensions: span 12.5m (40ft 11.5in); length 10.6m (34ft 9.5in); height 3.8m (12ft 7in); wing area 21.7m^2 (234sq ft)
Year entered service: 1944

Mikoyan-Gurevich MiG-3

Country of Origin: Soviet Union
Type: interceptor fighter

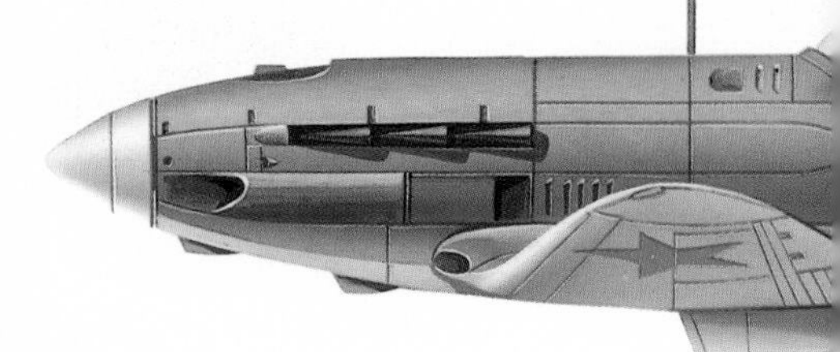

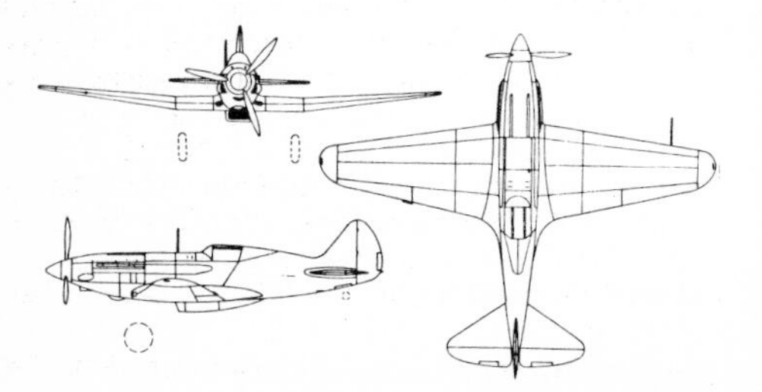

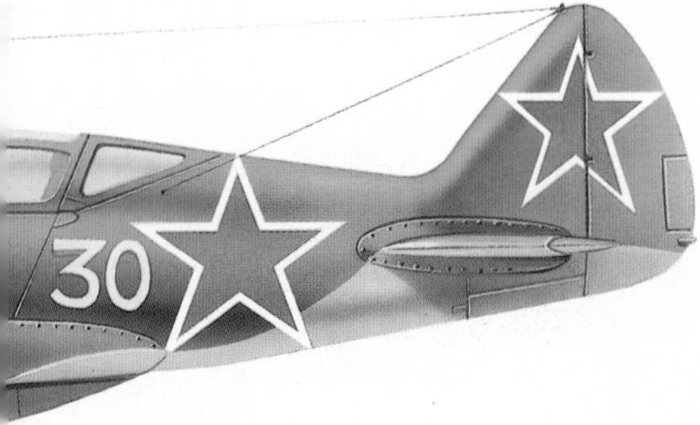

Accommodation: one
Armament: two 7.6mm ShKAS machine-guns plus one 13mm (.5in) Beresin BS machine-gun in upper cowling; provision for 200kg (440lb) bombs or six 82mm RS 82 rocket missiles underwing.
Power Plant: one 1,350hp Mikulin AM-35A
Performance: maximum speed 640km/h (398mph) at 7,800m (25,590 ft); service ceiling 12,000m (39,370ft); range 1,250km (777 miles)
Weights: empty 2,595kg (5, 721lb); loaded 3,350kg (7,385lb)
Dimensions: span 10.3m (33ft 9.5in); length 8.15m (26ft 9in); height 3.5m (11ft 6in); wing area 17.4m^2 (187.7sq ft)
Year entered service: 1941

Mitsubishi A6M5 Zero-Sen 'Zeke'

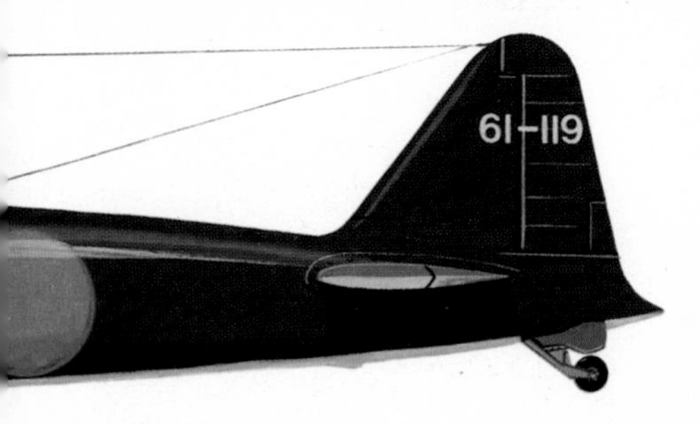

Country of Origin: Japan
Type: carrier-borne fighter
Accommodation: one
Armament: two 20mm Type 99 cannon in wings plus two 7.7mm Type 97 machine-guns in upper fuselage decking; bomb load 120kg (264lb)
Power Plant: one 1,130hp Nakajima NK1F Sakae 21
Performance: maximum speed 565km/h (351mph) at 6,000m (19,685ft); service ceiling 11,740m (38,520ft); maximum range 1,920km (1,194 miles)
Weights: empty 1,876kg (4,136lb); loaded 2,730kg (6,025lb)
Dimensions: span 11m (36ft 1in); length 9.12m (29ft 11in); height 3.51m (11ft 6.25in); wing area 21m^2 (229sq ft)
Year entered service: 1940

Mitsubishi G4M Model 22 'Betty'

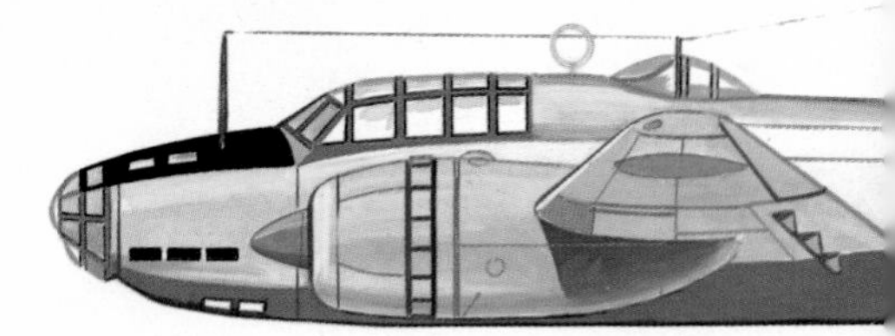

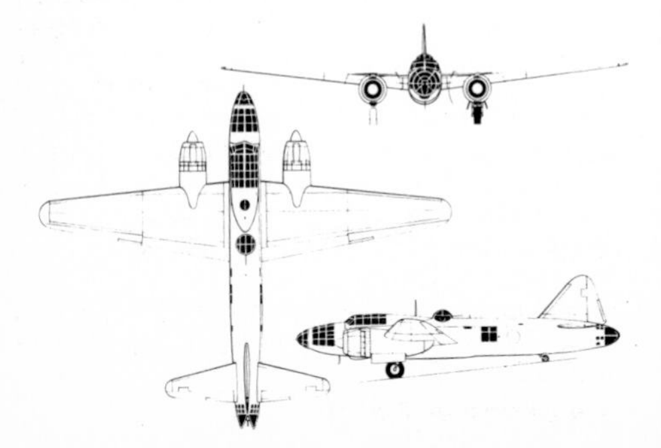

Country of Origin: Japan
Type: bomber
Accommodation: seven
Armament: two 7.7mm Type 92 machine-guns each in nose and fuselage beam positions; one 20mm Type 99 Model 1 cannon each in dorsal and tail turret; maximum bomb load 1,000kg (2,205lb) or one 800kg (1,764lb) torpedo
Power Plant: two 1,800hp Mitsubishi MK4P Kasei 21
Performance: maximum speed 438km/h (272mph) at 4,600m (15,090ft); service ceiling 8,950m (29,365ft); range 5,260km (3,270 miles)
Weights: empty 8,160kg (17,990lb); loaded 12,500kg (27,558lb)
Dimensions: span 25m (82ft .25in); length 20m (65ft 7.5in); height 6m (19ft 8.25in); wing area 78.1m^2 (840.9sq ft)
Year entered service: 1943

Nakajima Ki-43-1a Hayabusa 'Oscar'

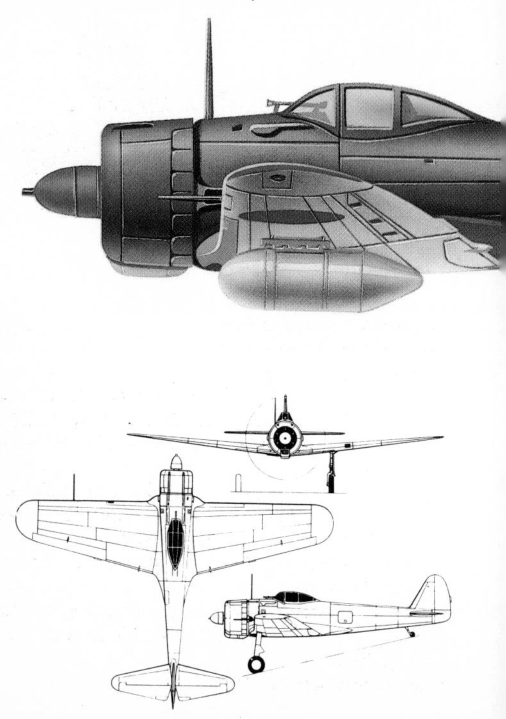

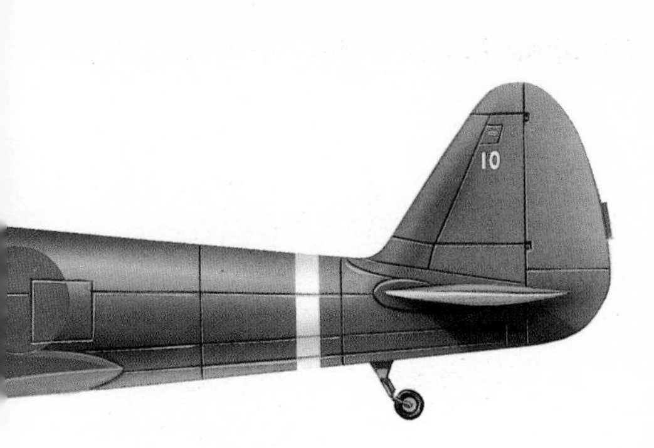

Country of Origin: Japan
Type: fighter/fighter-bomber
Accommodation: one
Armament: two 7.7mm Type 89 machine-guns in upper cowling; maximum bomb load 30kg (66lb)
Power Plant: one 980hp Nakajima Ha-25, Army Type 99
Performance: maximum speed 495km/h (308mph) at 4,000m (13,120ft); service ceiling 11,750m (38,500ft); range 1,200km (745 miles)
Weights: empty 1,580kg (3,483lb); maximum 2,583kg (5,695lb)
Dimensions: span 11.4m (37ft 6.25in); length 8.8m (28ft 11.75in); height 3.3m (10ft 8.75in); wing area 22m^2 (236.8sq ft)
Year entered service: 1941

North American B-25A Mitchell

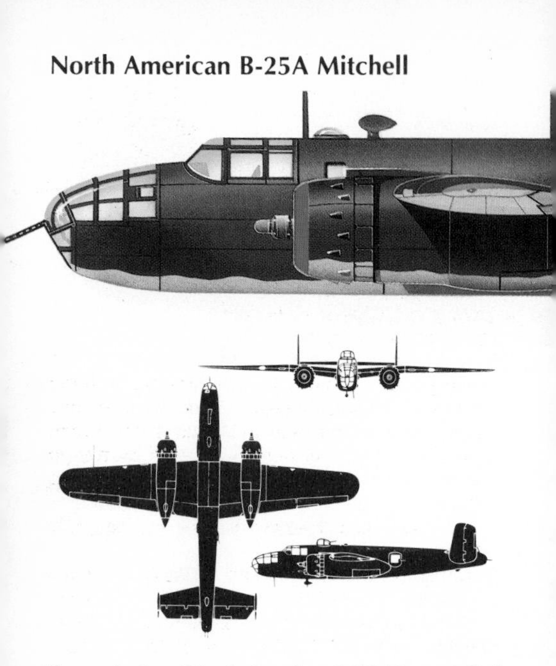

The much altered North-American B-25J-20 as supplied to the Soviet Union

Armament: two 13mm (0.5in) machine-guns in tail turret

Country of Origin: United States
Type: light day bomber
Accommodation: four–five
Armament: one 13mm (.5in) machine-gun in nose; two 13mm (.5in) machine-guns in dorsal turret; two 13mm (.5in) machine-guns in tail turret; one 13mm (.5in) machine-gun in ventral location or two in ventral turret; maximum bomb load 2,720kg (6,000lb)
Power Plant: two 1,350hp Wright Cyclone GR-2600A-5B
Performance: maximum speed 470km/h (292mph) at 4,572m (15,000ft); service ceiling 6,096m (20,000ft); maximum range 2,631km (1,635 miles)
Weights: empty 7,260kg (16,000lb); loaded 11,110kg (24,500lb)
Dimensions: span 20.6m (67ft 6.75in); length 16.5m (54ft 1in); height 4.8m (15ft 9.75in); 56.7m^2 (610sq ft)
Year entered service: 1940

North American P-51D Mustang

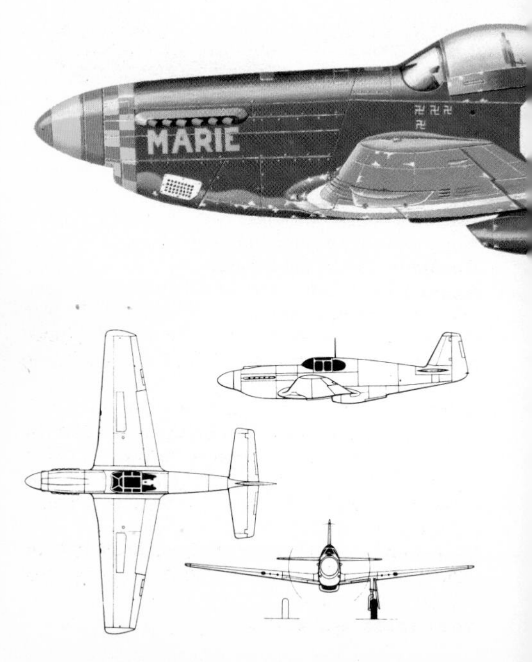

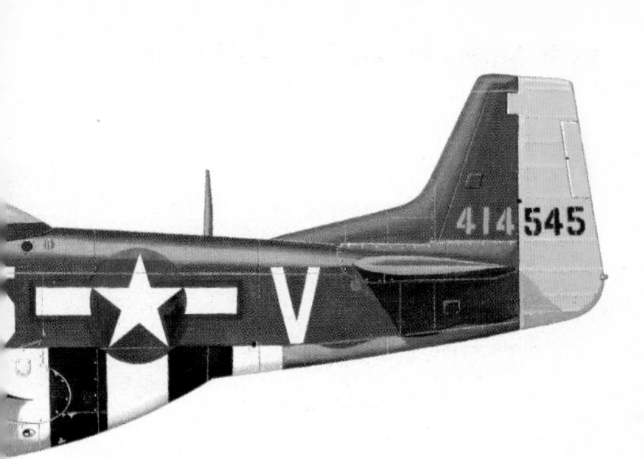

Country of Origin: United States
Type: fighter
Accommodation: one
Armament: six 13mm (.5in) Browning machine-guns; maximum bomb load 907kg (2,000lb), or six 13mm (.5in) rocket projectiles
Power Plant: one 1,510hp Packard Merlin V-1650-7
Performance: maximum speed 703km/h (437mph) at 7,620m (25,000ft); service ceiling 12,770m (41,900ft); maximum range 3,700km (2,300 miles).
Weights: empty 3,232kg (7,125lb); maximum 5,262kg (11,600lb)
Dimensions: span 11.9m (34ft .25in); length 9.85m (32ft 3.25in); height 4.2m (13ft 8in); wing area 212.6m^2 (233.2sq ft)
Year entered service: 1942

Northrop P-61B Black Widow

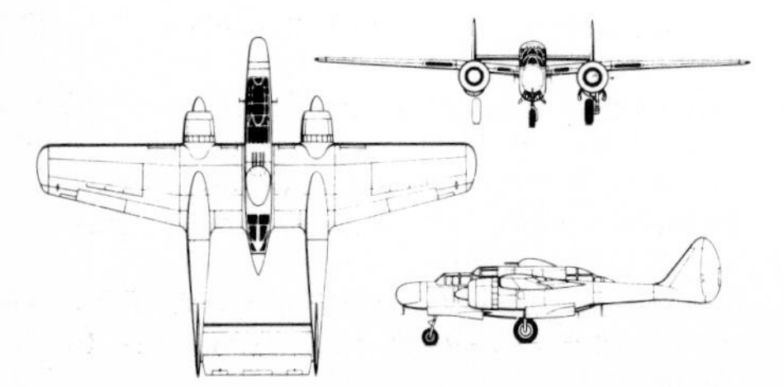

Northrop P-61A Black Widow

Country of Origin: United States
Type: night-fighter
Accommodation: three
Armament: four 13mm (.5in) guns; four 20mm cannon in remote-controlled dorsal turret; four 726kg (1,600lb) bombs underwing
Power Plant: two 2,000hp Pratt & Whitney R-2800-65
Performance: maximum speed 589km/h (366mph) at 6,096m (20,000ft); service ceiling 10,089m (33,100ft); maximum range 4,830km (3,000 miles)
Weights: empty 9,980kg (22,000lb); maximum 17,240kg (38,000lb)
Dimensions: span 20.1m (66ft); length 15.1m (49ft 7in) height 4.5m (14ft 8in); wing area 61.7m^2 (664sq ft)
Year entered service: 1944

Petlyakov Pe-2

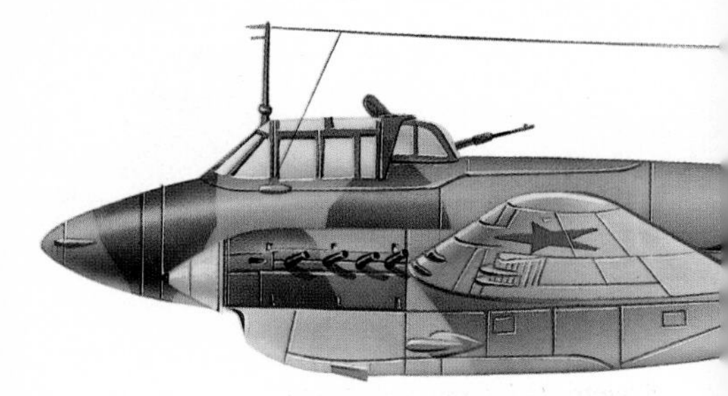

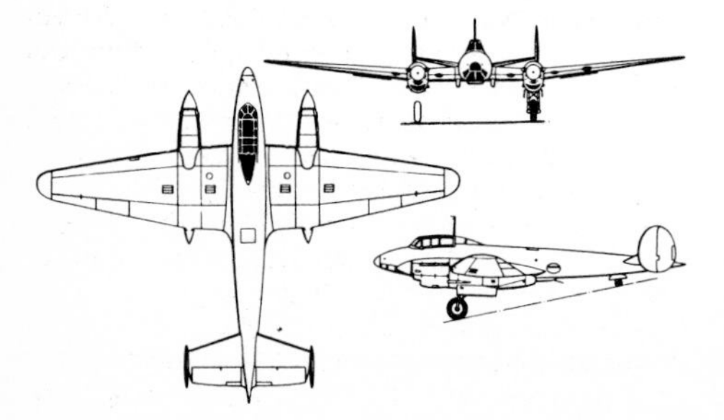

Country of Origin: Soviet Union
Type: dive-bomber
Accommodation: three–four
Armament: two 7.6mm ShKAS (or one 7.6mm ShKAS and one 12.7mm Beresin UBS) machine-gun in nose; plus one flexible 7.6mm ShKAS (or 12.7mm Beresin UBT) machine-gun each in dorsal or ventral position; maximum bomb load 1,200kg (2,645lb)
Power Plant: two 1,100hp Klimov M-105R
Performance: maximum speed 540km/h (336mph) at 5,000m (16,400ft); service ceiling 8,800m (28,900ft); range 1,500km (932 miles)
Weight: loaded 33,325kg (73,469lb)
Dimensions: span 17.2m (56ft 3.5in); length 12.6m (41ft 6.5in); height 4m (13ft 1.5in); wing area 40.5m^2 (436sq ft)
Year entered service: 1941

Republic P-47 Thunderbolt

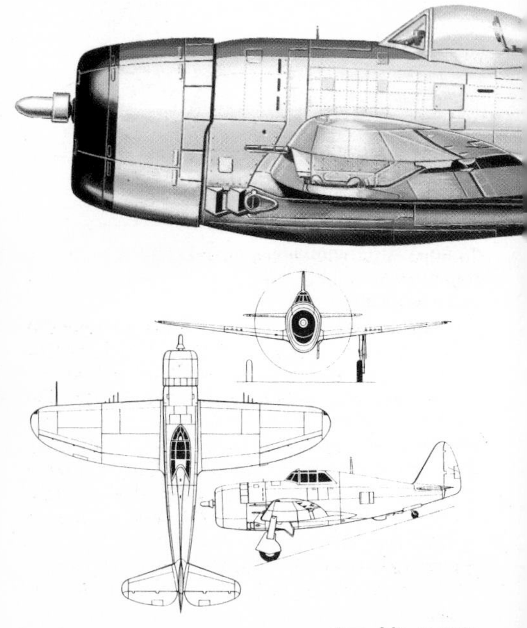

Republic P-47C

Country of Origin: United States
Type: fighter bomber
Accommodation: one
Armament: eight fixed 13mm (.5in) machine-guns in wings; maximum bomb load 907kg (2,000lb); six or eight 8cm (3in) rockets under wings if required
Power Plant: one 2,300hp Pratt & Whitney Double Wasp R-2800-59
Performance: maximum speed 687km/h (427mph) at 8,690m (28,500ft); service ceiling 11,278m (37,000ft); maximum range 3,170km (1,970 miles)
Weights: empty 4,540kg (10,000lb); loaded 6,630kg (14,600lb)
Dimensions: span 12.4m (40ft 9.25in); length 11m (36ft 1.75in); height 3.9m (12ft 7.75in); 27.9m^2 (300sq ft)
Year entered service: 1942

Savoia Marchetti S.M.79-1 Sparviero

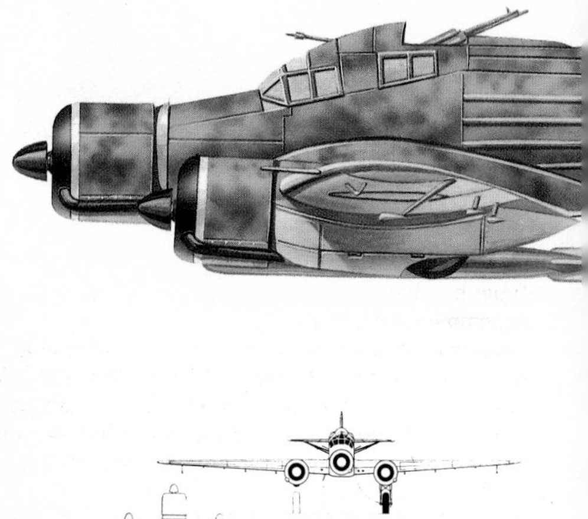

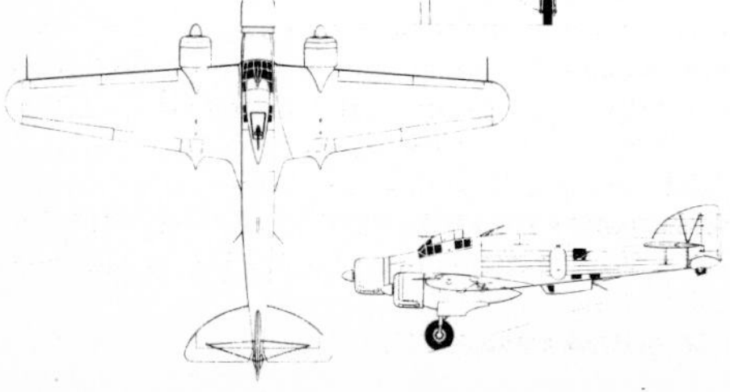

Country of Origin: Italy
Type: bomber
Accommodation: four–five
Armament: one 12.7mm Breda-SAFAT machine-gun above pilot's cockpit; one 12.7mm Breda-SAFAT machine-gun each in dorsal and ventral positions; one 7.7mm Lewis machine-gun in either of two lateral hatches; maximum bomb load 1,250kg (2,750lb)
Power Plant: three 780hp Alfa Romeo 126 RC 34
Performance: maximum speed 430km/h (267mph) at 4,000m (13,120ft); service ceiling 6,500m (21,320ft); normal range 1,900km (1,180 miles)
Weight: loaded 10,480kg (23,100lb)
Dimensions: span 21.2m (69ft 6.6in); length 15.8m (51ft 10in); height 4.3m (14ft 1.75in); wing area $61.7m^2$ (664.1sq ft)
Year entered service: 1937

Short Stirling I

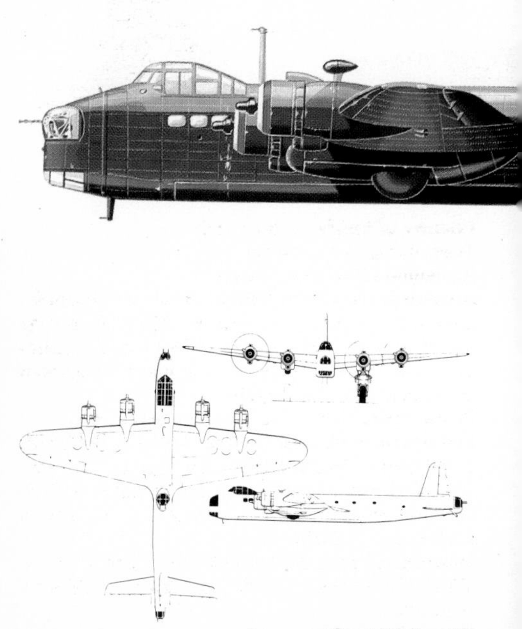

Short Stirling III

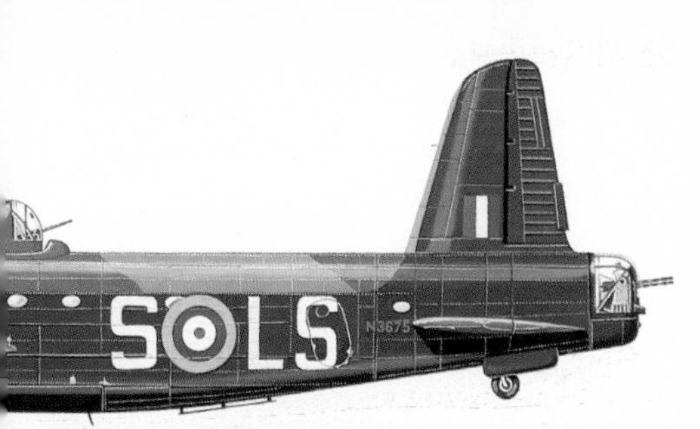

Country of Origin: Great Britain
Type: heavy night bomber
Accommodation: seven–eight
Armament: two 8mm (.303in) Browning machine-guns in nose turret; two 8mm (.303in) Browning machine-guns in dorsal turret; four 8mm (.303in) Browning machine-guns in tail turret; maximum bomb load 6,350kg (14,000lb)
Power Plant: four 1,595hp Bristol Hercules XI
Performance: maximum speed 434km/h (270mph) at 4,420m (14,500ft); service ceiling 5,490m (18,000ft); maximum range 3,240km (2,010 miles)
Weights: empty 21,200kg (46,900lb); maximum 31,790kg (70,000lb)
Dimensions: span 30.2m (99ft 1in); length 26.6m (87ft 3in); height 6.9m (22ft 9in); wing area 135.6m^2 (1,460sq ft)
Year entered service: 1940

Short Sunderland II

Country of Origin: Great Britain
Type: maritime reconnaissance and patrol flying boat

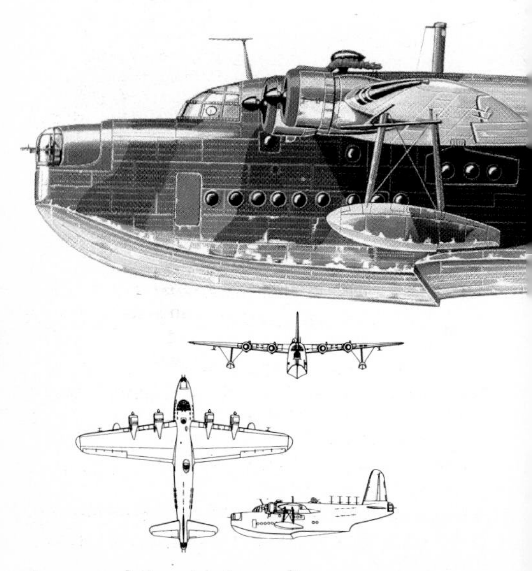

Accommodation: ten (normal)
Armament: four 8mm (.303in) Browning machine-

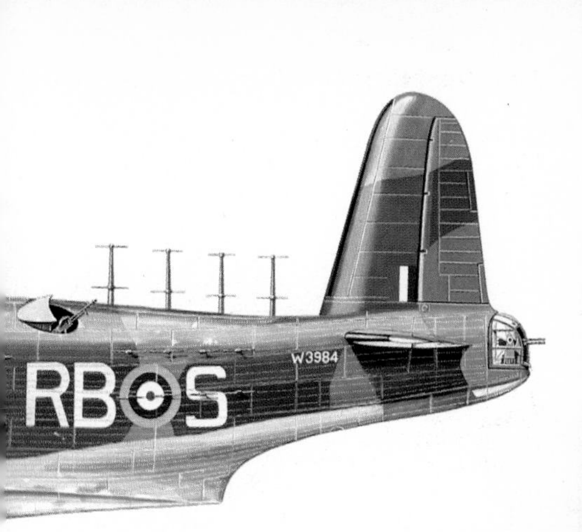

guns in nose and tail turrets; maximum bomb load 2,250kg (4,960lb)

Power Plant: four 1,010hp Bristol Pegasus XVIII

Performance: speed 343km/h (213mph); service ceiling 5,456m (17,900ft); maximum range 3,396km (2,110 miles)

Weights: empty 16,780kg (37,000lb); loaded 29,480kg (65,100lb)

Dimensions: span 34.4m (112ft 9.5in); length 26m (85ft 4in); height 10m (32ft 10.5in); 138m^2 (1,482sq ft)

Year entered service: 1938

Supermarine Spitfire Mk VA

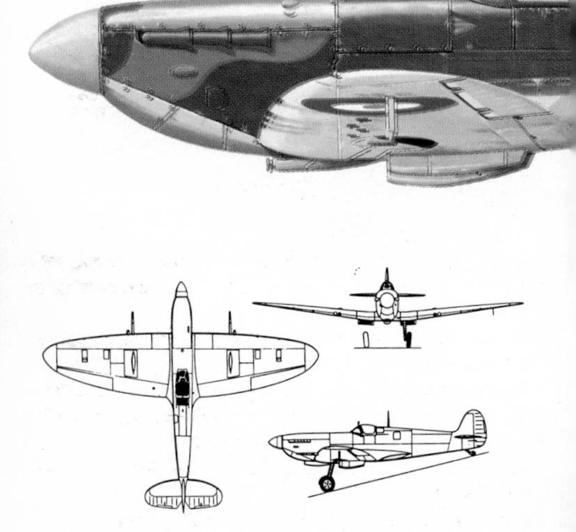

Country of Origin: Great Britain
Type: fighter
Accommodation: one
Armament: two 20mm Hispano cannon; four 8mm (.303in) Browning machine-guns; maximum bomb load 227kg (500lb)
Power Plant: one 1,470hp Rolls-Royce Merlin 45M, 50M or 55M
Performance: maximum speed 534km/h (332mph) at sea level; service ceiling 10,820m (35,500ft); maximum range 756km (470 miles)
Weights: empty 2,290kg (5,050lb); loaded 3,020kg (6,650lb)
Dimensions: wing span 9.8m (32ft 2in); length 9.1m (29ft 11in); height 3m (9ft 11in); wing area 21.5m^2 (231sq ft)
Year entered service: 1941

Vickers Wellington Ic

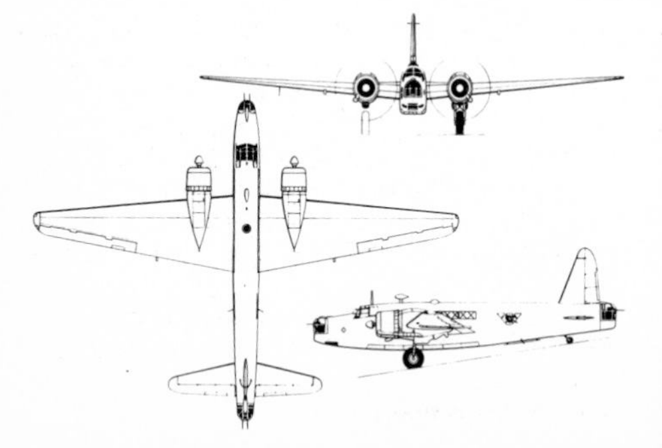

Country of Origin: Great Britain
Type: medium bomber
Accommodation: five–six
Armament: two 8mm (.303in) machine-guns in nose and tail turrets; two manually-operated beam guns; maximum bomb load of 2,040kg (4,500lb)
Power Plant: two 1,000hp Bristol Pegasus XVIII
Performance: maximum speed 378km/h (235mph) at 4,720m (15,500ft); service ceiling 5,490m (18,000ft); maximum range 4,100km (2,550 miles)
Weights: empty 8,420kg (18,556lb); loaded 12,930kg (28,500lb)
Dimensions: span 26.3m (86ft 2in); length 19.7m (64ft 7in); height 5.3m (17ft 5in); wing area 5.3m (17ft 5in)
Year entered service: 1938

Westland Lysander III

Country of Origin: Great Britain
Type: army co-operation
Accommodation: two

Armament: two 8mm (.303in) Browning machine-guns in wheel spats; two 8mm (.303in) Browning machine-guns in rear cockpit; sixteen 9kg (20lb) bombs or four 51kg (112lb) or 54kg (120lb) bombs, or two 113kg (250lb) on rear fuselage racks and detachable stub wing carriers
Power Plant: one 870hp Bristol Mercury XX
Performance: maximum speed 336km/h (209mph); service ceiling 6,553m (21,500ft); range 965km (600 miles)
Weights: empty 1,980kg (4,365lb); loaded 2,865kg (6,318lb)
Dimensions: span 15.2m (50ft); length 9.3m (30ft 6in); height 4.4m (14ft 6in); wing area 24.2m^2 (260sq ft)
Year entered service: 1938

Yakovlev Yak-1

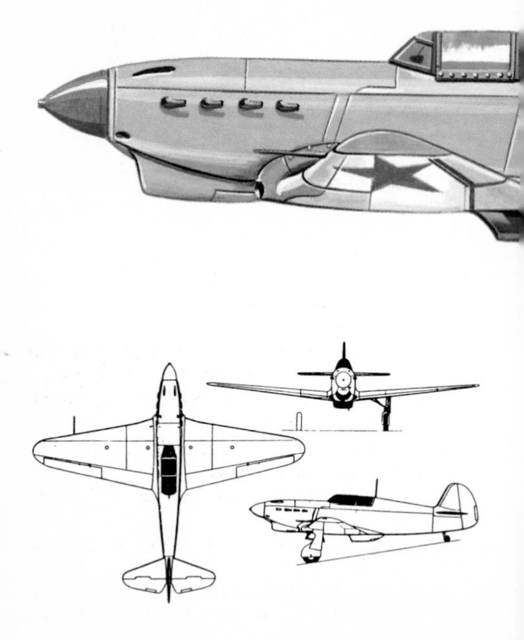

Country of Origin: Soviet Union
Type: fighter/fighter-bomber
Accommodation: one
Armament: one 20mm ShVAK cannon firing through propeller hub; two 7.6mm ShKAS machine-guns in upper cowling; maximum bomb load 200kg (440lb) or six 82mm RS 82 rocket projectiles
Power Plant: one 1,100hp Klimov M-105 PA
Performance: maximum speed 580km/h (360mph) at sea level; service ceiling 10,000m (32,810ft); range 850km (528 miles)
Weights: empty 2,330kg (5,137lb); loaded 2,820kg (6,217lb)
Dimensions: span 10m (32ft 9.75in); length 8.47m (27ft 9.5in); height 2.64m (8ft 8in); wing area 17.15m^2 (184.6sq ft)
Year entered service: 1940

Yakovlev Yak-9D

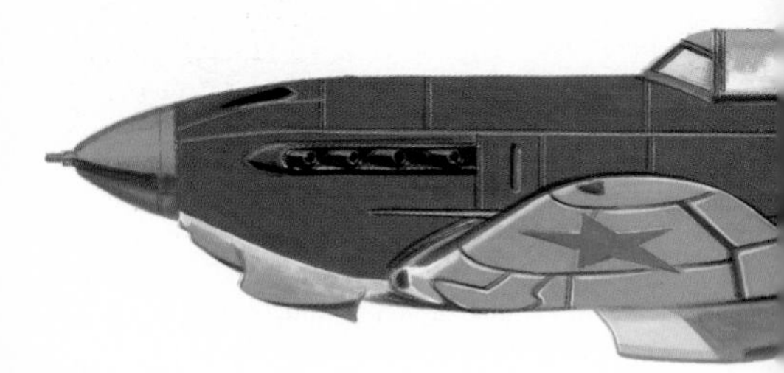

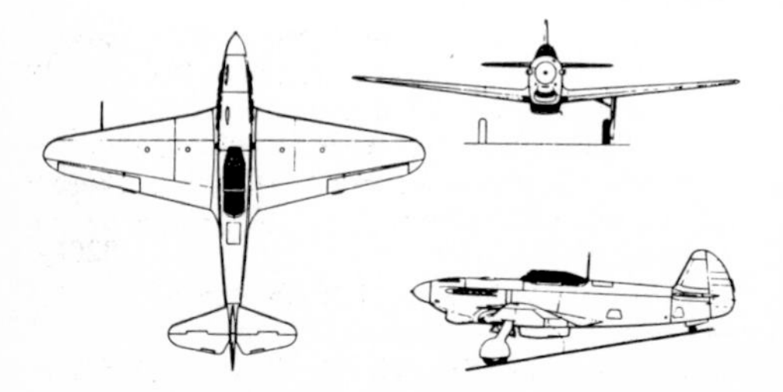

Country of Origin: Soviet Union
Type: long-range fighter
Accommodation: one
Armament: one 20mm ShVAK cannon firing through propeller hub, plus one 12.7mm Beresin UB machine-gun in upper cowling
Power Plant: one 1,360hp Klimov VK-105PF-3
Performance: maximum speed 545km/h (339mph); service ceiling 10,200m (33,464ft); range 825km (513 miles)
Weights: empty 2,480kg (5,467lb); loaded 3,010kg (6,636lb)
Dimensions: span 10m (32ft 9.75in); length 8.47m (27ft 9.5in); wing area 17.2m^2 (184.6sq ft)
Year entered service: 1942

Index

A6M5 (Mitsubishi) 204
A-20G (Douglas) 132
Aichi D3A1 'Val' 13, 92
Airco D.H.2 16
Albatros D.III 18
Albatros D.Va 6,20
Arado Ar 234B-2 Blitz 94
Armstrong Whitworth Whitley MkV 10, 96
Avenger (Grumman) 156
Avro Lancaster MkI 10, 98

B-10 (Martin) 192
B-17G (Boeing) 104
B-24D (Consolidated) 120
B-25 (North American) 210
B-26B (Martin) 194
B-29 (Boeing) 106
Battle (Fairey) 136
B.E.12 (Royal Aircraft Factory) 64
Bell P-39 Airacobra 100
Bf 109G-2 (Messerschmitt) 196
Bf 110F-2 Messerschmitt) 198
Black Widow (Northrop) 214
Blohm und Voss Bv 138A-1 102
Boeing B-17G Flying Fortress 13, 104
Boeing B-29 Superfortress 106
Boulton Paul Defiant MkII 11, 108
Bristol Beaufighter MkI 110
Bristol Beaufort 112
Bristol Blenheim IV 114
Bristol F.2B 22
Bristol Scout C 6, 24

C.V (Halberstadt) 38
C-47 (Douglas) 134
Camel (Sopwith) 78
Caproni Ca5 26
Chance-Vought F4U-1 Corsair 116
Consolidated B-24D Liberator 13, 120
Consolidated PBY-5 Catalina 13, 122
Curtiss P-40 Warhawk 124

D.II (Halberstadt) 40
D3A1 (Aichi) 92
D.III (Albatros) 18
D.III (Pfalz) 62
D.III (Siemens-Schuckert) 70
D.Va (Albatros) 20

D.VII (Fokker) 28
D.VIII (Fokker) 30
Defiant (Boulton Paul) 108
De Havilland D.H.9 26
De Havilland Mosquito MkXVI 5, 13, 126
D.H.2 (Airco) 16
Dolphin (Sopwith) 82
Dornier Do 12Z 8, 128
Dornier Do 217K 10, 130
Douglas A-20G 13, 132
Douglas C-47 134
Dr.I (Fokker) 32

E.III (Fokker) 34

F.2B (Bristol) 22
F4F (Grumman) 152
F4U-1 (Chance-Vought) 118
F6F (Grumman) 154
Fairey Batle 136
Fairey Swordfish 138
Fiat G.50bis Freccia 12, 140
Flying Fortress (Boeing) 104
Focke-Wulf Fw 189A-1 Uhu 142
Focke-Wulf Fw 190 14, 144
Focke-Wulf 200C 146
Fokker D.VII 7, 28
Fokker D.VIII 30
Fokker Dr.I 32
Fokker E.III 6, 34
Fw 189A-1 (Focke-Wulf) 142
Fw 190 (Focke-Wulf) 144
Fw 200C (Focke-Wulf) 146

G4M (Mitsubishi) 206
G.V (Gotha) 36
G.50bis (Fiat) 140
G.102 (Martinsyde) 50
Gloster Gladiator 148
Gloster Meteor I 15, 150
Gotha G.V 7, 36
Grumman F4F Wildcat 12, 152
Grumman F6F Hellcat 12, 154
Grumman TBF Avenger 12, 156

Halberstadt C.V 38
Halberstadt D.II 40
Handley Page O/400 7, 42
Handley Page V/1500 44
Handley Page Halifax 10, 158
Handley Page Hampden 10, 160
Hansa-Brandenburg

W.12 46
Hansa-Brandenburg
W.29 48
Hawker Hurricane I 8, 162
Hawker Tempest 164
Hawker Typhoon 166
Heinkel He111 8, 168
Heinkel He219A-5/R2
Uhu 170
Hellcat (Grumman) 154
Hurricane (Hawker) 162

Ilya Mouromets (Sikorsky) 76
Ilyushin Il-2m3 5, 172
Ilyushin Il-10 12, 174

Junkers Ju 52/3m g7e 176
Junkers Ju 87B-1 178
Junkers Ju 88D-1 180

Lancaster (Avro) 98
Lavochkin La-7 184
Lavochkin LaGG-3 182
Liberator (Consolidated) 122
Lockheed P-38 Lightning 186
Lysander (Westland) 230

Macchi MC.200 Saetta 12, 188
Macchi MC.205 Veltro 190
Martin B-10 192
Martin B-26B Marauder 13, 194
Martinsyde G. 100
Elephant 50
Messerschmitt Bf 109G-2 8, 196
Messerschmitt Bf 110-2 11, 198
Messerschmitt Me 262
A-1a 200
Meteor (Gloster) 150
Mikoyan-Gurevich MiG-3 12, 202
Mitchell (North
American) 210
Mitsubishi A6M5 Model-52
Reisen 'Zeke' 12, 204
Mitsubishi G4M Model 22
'Betty' 206
Morane-Saulnier Type L 52
Morane-Saulnier Type N 54
Mosquito (De Havilland) 126

Nakajima Ki-43-1a
Hayabusa 'Oscar' 208
Nieuport XII 56
Nieuport XVII 58
Nieuport Scout XXVII 6, 60
North American B-25

Mitchell 13, 210
North American P-51D Mustang 13, 212
Northrop P-61 Black Widow 214

O/400 (Handley Page) 42

P-38 (Lockheed) 186
P-39 (Bell) 100
P-40 (Curtiss) 124
P-47 (Republic) 218
P-51D (North American) 212
P-61B (Northrop) 214
PBY-5 (Consolidated) 122
Petlyakov Pe-2 216
Pfalz D.III 6, 60
Pup (Sopwith) 82

R1 (Siemens-Schuckert) 72 74
Republic P-47 Thunderbolt 13, 218
Royal Aircraft Factory B.E.12 6, 64
Royal Aircraft Factory S.E.2b 66
Royal Aircraft Factory S.E.5 68
Savioa Marchetti S.M.79-1 Sparviero 12, 220
Scout XXVII (Nieuport) 60
Scout (Bristol) 24
S.E.2b (Royal Aircraft Factory) 66
S.E.5 (Royal Aircraft Factory) 68
Short Stirling I 10, 222
Short Sunderland II 13, 224
Siemens-Schuckert D.III 70
Siemens-Schukert R1 72
Sikorsky Ilya Mouromets 74
S.M.79-1 (Savoia-Marchetti) 220
Sopwith 1½ Strutter 76
Sopwith Camel F.1 7, 78
Sopwith Dolphin Mk.1 7, 80
Sopwith Pup 6, 82
Sopwith Triplane 86
Spad VII 78
Spad XIII 6, 90
Spitfire (Supermarine) 226
Strutter (Sopwith) 1½ 76
Sunderland (Short) 224
Superfortress (Boeing) 108
Supermarine Spitfire Mk Va 8, 226
Swordfish (Fairey) 138
Tempest (Hawker) 164
Triplane (Sopwith) 86
Type L (Morane-Saulnier) 52
Type N (Morane-Saulnier) 54
Typhoon (Hawker) 166

V/1500 (Handley Page) 44
Vickers Wellington K 10, 228
Westland Lysander III 5, 230
Yakovlev Yak-1 12, 232
Yakovlev Yak-9D 234
Zero (Mitsubishi) 204